Principles of
Engineering Mechanics

Principles of Engineering Mechanics

Contributors

Mohamed El Naschie et al.

AURIS
Reference

www.aurisreference.com

Principles of Engineering Mechanics

Contributors: Mohamed El Naschie et al.

Published by Auris Reference Limited
www.aurisreference.com

United Kingdom

Principles of Engineering Mechanics

ISBN: 978-1-78154-815-8

British Library Cataloguing in Publication Data
A CIP record for this book is available from the British Library

Printed in the United Kingdom

Exclusively distributed by CBS Publishers & Distributors Pvt. Ltd.

Sales & Distribution Rights only for India, Pakistan, Bangladesh, Sri Lanka, Nepal and Bhutan. This book is not to be sold outside these territories.

Contents

List of Abbreviations

VPM	Virtual physical model
DRS	Drag reduction system
ARGAs	Adaptive genetic algorithm
BLDC	Brushless direct current
RGA	Real-coded genetic algorithm
SRGA	Standard genetic algorithm
DS	Drill string
MSE	Mechanical and systems engineering

List of Contributors

Mohamed El Naschie
Department of Physics, Faculty of Science, University of Alexandria, Alexandria, Egypt

Alice Rosa da Silva
School of Civil Engineering, Federal University of Uberlândia, Uberlândia, Brazil

Aristeu da Silveira-Neto
School of Mechanical Engineering, Federal University of Uberlândia, Brazil

Antônio Marcos Gonçalves de Lima
School of Mechanical Engineering, Federal University of Uberlândia, Brazil

Roberto Capata
Department of Mechanical and Aerospace Engineering, University of Roma "Sapienza", Rome, Italy

Leone Martellucci
Department of Mechanical and Aerospace Engineering, University of Roma "Sapienza", Rome, Italy

Rong-Fong Fung
Department of Mechanical & Automation Engineering, National Kaohsiung First University of Science and Technology, Kaohsiung, Taiwan

Chun-Hung Lin
Graduate Institute of Electrical Engineering, National Kaohsiung First University of Science and Technology, Kaohsiung, Taiwan

Nabil W. Musa
Department of Mechanical Engineering, Philadelphia University, Amman, Jordan

V. I. Gulyayev
Department of Mathematics, National Transport University, Kyiv, Ukraine

L. V. Shevchuk
Department of Mathematics, National Transport University, Kyiv, Ukraine

Hasan Aldabas
Department of Mechanical Engineering, Philadelphia University, Amman, Jordan

Paul Ndy Von Kluge
Department of Physics, Mechanics Laboratory, Faculty of Sciences, University of Yaoundé 1, Yaoundé, Cameroon

Djuidjé Kenmoé Germaine
Department of Physics, Mechanics Laboratory, Faculty of Sciences, University of Yaoundé 1, Yaoundé, Cameroon

Kofané Timoléon Crépin
Department of Physics, Mechanics Laboratory, Faculty of Sciences, University of Yaoundé 1, Yaoundé, Cameroon

Berend Denkena
Leibniz Universität Hannover, Institute of Production Engineering and Machine Tools, An der Universität 2, Hanover, Germany

Martin Eckl
Leibniz Universität Hannover, Institute of Production Engineering and Machine Tools, An der Universität 2, Hanover, Germany

Dominik Brouwer
Leibniz Universität Hannover, Institute of Production Engineering and Machine Tools, An der Universität 2, Hanover, Germany

Jakkula Anand Rao
Department of Mathematics, Osmania University, Hyderabad, Telangana

Gandamalla Vasumathi
Department of Mathematics, Osmania University, Hyderabad, Telangana

Jakkula Mounica
Department of Mathematics, NGRI, Hyderabad, Telangana

Tom Joyce
School of Mechanical and Systems Engineering, Newcastle University, Stephenson Building, Claremont Road, Newcastle upon Tyne NE1 7RU, UK

Iain Evans
School of Mechanical and Systems Engineering, Newcastle University, Stephenson Building, Claremont Road, Newcastle upon Tyne NE1 7RU, UK

William Pallan
WP Consulting, Newcastle upon Tyne, UK

Clare Hopkins
School of Mechanical and Systems Engineering, Newcastle University, Stephenson Building, Claremont Road, Newcastle upon Tyne NE1 7RU, UK

Zhenghao Liu
College of Mechatronics and Control Engineering, Shenzhen University, Shenzhen, China

Hong Chen
College of Mechatronics and Control Engineering, Shenzhen University, Shenzhen, China

Wael N. Abd Elsamee
Faculty of Engineering, Sinai University, El Arish, Egypt

Tawfik El-Sayed Tawfik
Department of Mathematics, Faculty of Science, Mansoura University, Mansoura, Egypt

Preface

Engineering mechanics is the application of mechanics to solve problems involving common engineering elements. The book Principles of Engineering Mechanics provides the basis for a stimulating and rewarding one-term course for advanced undergraduate and first-year graduate students specializing in mechanics, engineering science, engineering physics, applied mathematics, materials science, and mechanical, aerospace, and civil engineering. First chapter focuses on cosmic dark energy density from classical mechanics and seemingly redundant Riemannian finitely many tensor components of Einstein's general relativity. Second chapter highlights on rotational oscillation effect on flow characteristics of a circular cylinder at low Reynolds number. Third chapter gives an approach on aerodynamic brake for formula cars. In fourth chapter, the main objective is to identify the parameters of motors, which includes a brushless direct current (BLDC) motor and an induction motor. The motor systems are dynamically formulated by the mechanical and electrical equations. Fifth chapter deals with the theoretic simulation of a drill bit whirling under conditions of its contact interaction with the bore-hole bottom rock plane. Sixth chapter emphasizes on dry friction with various frictions laws. Seventh chapter presents a novel design for a knee disarticulation prosthesis. In this design, three hydraulic cylinders form the supporting structure and provide the damping effect at the same time. A numerical study on boundary layer flow behavior, heat and mass transfer characteristics of a nanofluid over an exponentially stretching sheet in a porous medium is presented in eighth chapter. Ninth chapter presents a hands-on project-based mechanical engineering design module focusing on sustainability. An automated assembly system is studied in tenth chapter and a new method is proposed in eleventh chapter to calculate the ultimate capacity of the pile from pile load test data. In twelfth chapter, two control torques which stabilize asymptotically the rotational motion of an axi-symmetric rigid body are obtained only in terms of the orientation parameters.

Chapter 1

COSMIC DARK ENERGY DENSITY FROM CLASSICAL MECHANICS AND SEEMINGLY REDUNDANT RIEMANNIAN FINITELY MANY TENSOR COMPONENTS OF EINSTEIN'S GENERAL RELATIVITY

Mohamed El Naschie

Department of Physics, Faculty of Science, University of Alexandria, Alexandria, Egypt

ABSTRACT

We determine the limit of the ratio formed by the independent components of the Riemann tensor to the non-zero component as space dimensionality tends to infinity and find it to be 12. Subsequently we use this result in conjunction with Newtonian classical mechanics to show that the ordinary measurable cosmic energy density is given by $E(O) = \left(\frac{1}{2}\right)\left(\frac{1}{12-1}\right)mc^2$ while the dark energy density is obviously the Legendre transformation dual energy E(D) = 1 ⊡ E(O). The result is in complete agreement with the COBE, WMAP and type 1a supernova measurements.

INTRODUCTION

The present analysis, although quite short and straight forward, is never the less of considerable importance to the understanding of arguably one of the most difficult questions presently facing both theoretical physics and cosmology, i.e., the meaning and nature of dark energy [1] -[8] . Unlike previous publications we concentrate here on taking the limit of infinite dimensional geometry and proceed from there, using Newtonian mechanics only, to show that the analysis and results are consistent with a physical truly fractal and Cantorian spacetime [4] -[10] . Finally we hint at a connection to nonlocal elasticity [11] -[13] .

ANALYSIS

In the following short communication we employ what is essentially a fractal spacetime picture to compute the dark energy density [1] -[10] via the ratio of non-zero components [1] to the total number of independent components in n dimensions when we let n tend to infinity. We recall that voids in spacetime are essentially empty sets with a negative Menger-Uhryson topological dimension while ordinary spacetime are zero sets [10]. Said in the language of fractals, empty regions are fat fractals while non-empty regions of spacetime are thin fractals [4] -[10]. Translated to the algebraic geometry of a Riemannian tensor, the ordinary spacetime regions are represented by the non-zero component of the tensor in n dimensions which are given as is well known by [1] :

$$N_1 = N\left(\text{non-zero component}\right) = n^2\left(n-1\right)^2 \tag{1}$$

On the other hand the total numbers of independent components are given by the well known Formula:

$$N_2 = N\left(\text{independent component}\right) = n^2\left(n^2-1\right)/12 \tag{2}$$

Following what we reasoned earlier on then we can regard the sparsity ratio N_2/N_1 as essentially the Lorentzian factor in Einstein's Formula [7] -[10]. That means we regard $E = \gamma mc^2 = mc^2$ as the regular hundred percent total energy density which is a basic tacit assumption in all our non-fractal theories in high energy physics and cosmology. Consequently Newton's kinetic energy [4] -[10] :

$$E_N = \frac{1}{2}mv^2 \tag{3}$$

could be thought of as a generalization of Einstein's energy when $v \to c$ and $\frac{1}{2} \to 1$ [10]. This limit corresponds in set theoretical formulation to when the zero set $d_c^{(0)} = \phi$ changes its average value $\left\langle d_c^{(0)}\right\rangle = \frac{1}{2}$ and tends to the unitary set $\left\langle d_c^{(0)}\right\rangle = 1$ [10]. Following this transfinite set theoretical logic we can write [5] -[8] :

$$E = \left(\gamma\right)\left[\frac{1}{2}m\left(v \to c\right)^2\right] = \left(\gamma/2\right)mc^2 \tag{4}$$

On the other hand for a Cantorian fractal formal dimensionality is intrinsically $n \to \infty$. Therefore by letting n go to infinity and using the role of de L'Hopital we have [1] :

$$\gamma(n \to \infty) = \left[n^2 (n^2 - 1)/12 \right] / \left[n^2 (n-1)^2 \right] = \frac{1}{12} \left[\frac{(n-1)(n+1)}{(n-1)^2} \right] = \frac{1}{12} \frac{n+1}{n-1} = \frac{1}{12}$$

(5)

Inserting in E we find the following slightly inaccurate result as will be explained later [4] -[10] :

$$E = \left(\frac{1}{2} \right) \frac{1}{12} m(v \to c)^2$$

(6)

Now a direct comparison between $E = mc^2$ and $E(0)$ could not be made unless we become completely aware that $E = mc^2$ was tacitly based on a one degree of freedom Lagrangian which Einstein never wrote down. This is of course the U(1) single photon degree of freedom leading to $E = mc^2$. The above formula however has taken on board also indirectly 12 degrees of freedom $\lambda = 1/\gamma = 12$ representing the 12 messenger photons of the standard model, i.e., $|SU(3)| = 8$ gluons, $|SU(2)| = 3$ (w^+, W^+, Z^o) and $|(U)| = 1$ photon [10] . Consequently consistency requires us to subtract the ordinary photon from our 12 messenger particles so that we do not double count it. In other words we have the following consistent limits:

$$d_c^{(0)} = \phi \to \left\langle d_c^{(0)} \right\rangle = \frac{1}{2}$$

(7)

$$\left\langle d_c^{(0)} \right\rangle = \frac{1}{2} \to d_c^{(0)} = d_c^{(1)} = 1$$

(8)

$$v \to c$$

(9)

$$\frac{1}{12} \to \frac{1}{12-1} = \frac{1}{11}$$

(10)

$$\gamma = 1 \to \gamma = \left(\frac{1}{2} \right) \left(\frac{1}{11} \right) = \frac{1}{22}$$

(11)

That way we find the correction $\gamma = \left(\frac{1}{2} \right) \left(\frac{1}{12-1} \right)$ and thus:

$$E = mc^2 \to E(0) = mc^2/22$$

(12)

From the above the dark energy density is easily deduced to be [4] -[10] :

$$E(D) = mc^2 - (mc^2/22) = mc^2 (21/22)$$

(13)

which is 95.5% of the total energy in full agreement with measurements as well as previous theoretical derivations. We could validate the preceding result in virtually one line of analysis when we consider that the only non-zero components of the Riemann tensor of general relativity is 24 [1] and compare that with the maximal possible of symmetries as given by the killing vector fields of Witten's 5 Brane theory in eleven dimensions, namely 528. The ratio γ is then given in a trivial straight forward way by [4] -[10] :

$$\gamma = \frac{24}{528} = \frac{1}{22}$$

(14)

exactly as shown above. Dark energy density is consequently [5] -[8] :

$$E(D) = mc^2 \frac{528-24}{528} = \frac{504}{528} mc^2 = mc^2 (21/22)$$

(15)

also as shown above.

CONCLUSION

In the present short analysis we started from an infinite dimensional geometry and topology and ended with a finite expression for both the ordinary and the dark energy density of the cosmos. The result, as expected, is in full agreement with measurements but we gained additional evidence for the true physical fractality of spacetime at both the quantum as well as the cosmological level[1] -[10] . It is crucial to appreciate the deep meaning and implications of obtaining a quantum gravity result in agreement with cosmic measurements using essentially Newtonian classical mechanics and without involving quantum mechanics. We stress that this is only possible because of our use of nonclassical fractal-Cantorian spacetime geometry and infinite dimensional topology[4] -[10] . We note on passing that we have very recently discovered that all the above results could also be obtained using the nonlocal elasticity theory of A. C. Eringen [6] [11] -[14] which is a welcome confirmation of the basic philosophy of the present work, namely the unity of physics, mathematics, engineering and philosophy [14] .

REFERENCES

1.   Martin, J.L. (1995) General Relativity. Prentice Hall, London. in particular pages 89 and 111.

2.   Linder, E. (2008) Dark Energy. Scholarpedia, 3, 4900.http://dx.doi.org/10.4249/scholarpedia.4900

3.   Caroll, S. (2013) Why Does Dark Energy Make the Universe Accelerate?http://www.preposterousuniverse.com/blog/2013/11/16/why-does-dark-energy-make-t

4.   El Naschie, M.S. (2014) Calculating the Exact Experimental Density of the Dark Energy in the Cosmos Assuming a Fractal Speed of Light. International Journal of Modern Nonlinear Theory and Application. in

press.

5. El Naschie, M.S. (2014) Cosserat-Cartan and de Sitter-Witten Spacetime Setting for Dark Energy. Quantum Matter. in press.

6. El Naschie, M.S. (2014) Pinched Material Einstein Spacetime Produces Accelerated Cosmic Expansion. International Journal of Astronomy and Astrophysics, 4, 80-90.http://dx.doi.org/10.4236/ijaa.2014.41009

7. El Naschie, M.S. (2013) the Quantum Gravity Immirzi Parameter—A General Physical and Topological in Terpretation. Gravitation and Cosmology, 19, 151-155.http://dx.doi.org/10.1134/S0202289313030031

8. El Naschie, M.S. (2014) Dark energy via Quantum Field Theory in Curved Spacetime. Journal of Modern Physics and Applications, 2, 1-7.

9. El Naschie, M.S. (2013) Experimentally Based Theoretical Arguments that Unruh's Temperature, Hawking's Vacuum Fluctuation and Rindler's Wedge Are Physically Real. American Journal of Modern Physics, 2, 357-361.http://dx.doi.org/10.11648/j.ajmp.20130206.23

10. El Naschie, M.S. (2013) Quantum Entanglement: Where Dark Energy and Negative Gravity Plus Accelerated Expansion of the Universe Comes from. Journal of Quantum Information Science, 3, 57-77. http://dx.doi.org/10.4236/jqis.2013.32011

11. Eringen, A.C. and Edelen, D.G. (1972) On Nonlocal Elasticity. International Journal of Engineering Science, 10, 233-248. http://dx.doi.org/10.1016/0020-7225(72)90039-0

12. Challamel, N., Wang, C.M. and Elishakoff, I. (2014) Discrete Systems Behave as Nonlocal Structural Elements; Bending Buckling and Vibration Analysis. European Journal of Mechanics—A/Solids, 44, 125-135. http://dx.doi.org/10.1016/j.euromechsol.2013.10.007

13. El Naschie, M.S. (1990) Stress, Stability and Chaos in Structural Engineering: An Energy Approach. McGraw Hill, London.

14. El Naschie, M.S. (2014) the Meta Energy of Dark Energy. Open Journal of Philosophy. In press.

Chapter 2

ROTATIONAL OSCILLATION EFFECT ON FLOW CHARACTERISTICS OF A CIRCULAR CYLINDER AT LOW REYNOLDS NUMBER

Alice Rosa da Silva[1], Aristeu da Silveira-Neto[2], and Antônio Marcos Gonçalves de Lima[2]

[1]School of Civil Engineering, Federal University of Uberlândia, Uberlândia, Brazil

[2]School of Mechanical Engineering, Federal University of Uberlândia, Brazil

ABSTRACT

Two dimensional numerical simulations of flow around a rotationally oscillating circular cylinder were performed at Re = 1000. A wide range of forcing frequencies, f_r, and three values of oscillation amplitudes, A, are considered. Different vortex shedding modes are observed for a fixed A at several values of f_r, as well as for a fixed f_r at different values of A. The 2C mode of vortex shedding was obtained in the present study. It is important to point out that this mode has not been observed by other investigators for rotationally oscillating case. Also, it is verified that this mechanism has great influence on the drag coefficient for high frequency values. Furthermore, the lift and pressure coefficients and the power spectra density are also analyzed.

INTRODUCTION

It is widely known that the vibrations induced by vortex shedding process may have negative effects in engineering systems, such as economic loss, damage of installations, and very frequently with environment-related consequences. Thus, although the fluid flow around circular cylinder is a classical problem in fluid dynamic due to its simple geometry, this is a reason for which in the last decades, a great deal of effort has been devoted to the development of numerical and computational procedures for dealing with the problem of vortex shedding phenomena. Comprehensive studies on this subject have been reported in books [1] -[3] .

In applications in which a circular cylinder under rotational oscillations is involved, the flow dynamic is different from those observed for stationary cylinders and has fascinated researchers for a long time [4] - [7] . In this case, two important parameters related to the prescribed motion are the forcing frequency and amplitude. Hence, the characterization of those parameters on the flow structure becomes of capital importance.

Among some studies over rotationally oscillating cylinders, the so-named Hybrid Vortex Method and the Discrete Vortex Method have been proposed [8] [9] to investigate the process of vortex formation for a set of forcing frequency at Reynolds number 200 and 1000. Srinivas and Fujisawa [10] have used the unsteady form of Reynolds-averaged Navier-Stokes equations combined with the k-e model of turbulence in order to determine the effects of several parameters on the flow structure over a rotationally oscillating circular cylinder for Reynolds in the range 2000 - 3.0×10^4. Ray and Christofides [11] studied a control system based on the open-loop simulations to reduce the effects of drag exerted on a circular cylinder subjected to rotational oscillations for Reynolds number in the range 100 - 500.

The present study focuses on two-dimensional, incompressible viscous flow over a rotationally oscillating circular cylinder by using the Immersed Boundary Methodology (IBM) [12] in order to investigate the oscillation effect in the wake structure behind the cylinder, at different forcing frequencies and amplitudes. In addition, the influence of those parameters on the drag, lift and pressure coefficients and the power spectra density are also analyzed. The simulations were carried out for flows at Re = 1000, amplitudes equals to 1, 2 and 3 and for various frequency ratios. The used methodology has showed a promising tool to simulate mobile bodies and in the present study, it captures the vortex shedding mode, named 2C, which has not been found for the case of rotational oscillation by other researchers.

GENERAL ASPECTS OF THE NUMERICAL METHODOLOGY

Mathematical Formulation for the Fluid and for the Fluid-Solid Interface

One approach, which is not very common to solve the Navier-Stokes equations, is the so called velocity-vorticity formulation [13] . Flows can also be modeled by these equations in primitive variables [14] . In this section, the IBM combined with the Virtual Physical Model (VPM) [15] is summarized. For a viscous and incompressible flow, the IBM is based on the Navier-Stokes

Equation (1) with an added force source term which acts on the fluid so that a particle perceives the existence of the solid interface. The continuity Equation (2) is also expressed.

$$\frac{\partial u_i}{\partial t} + \frac{\partial (u_i u_j)}{\partial x_j} = -\frac{1}{\rho}\frac{\partial p}{\partial x_i} + \frac{\partial}{\partial x_j}\left[v\left(\frac{\partial u_i}{\partial x_j} + \frac{\partial u_j}{\partial x_i}\right)\right] + f_i$$

(1)

$$\frac{\partial u_i}{\partial x_i} = 0$$

(2)

where ρ [kg/m³] and v [m²/s] are the specific mass and the kinematic viscosity, respectively; u_i [m/s] and p [N/m²] are, respectively, the i-th velocity component and the pressure; f_i [N/m³] is the i-th component of the Eulerian force calculated as follows:

$$f(x) = \sum_k D_{ij}(x - x_k)F(x_k)\Delta S^2(x_k)$$

(3)

In Equation (3) x [m] and x_k [m] are the position vectors of a Eulerian and a Lagrangian point, respectively, and ΔS [m] is the arc length centered in each Lagrangian point, which is evenly spaced as shown in Figure 1(a). $F(x_k)$ [N] is the Lagrangian force over the interface, and D_{ij} [m⁻²] is the distribution function [16] .

At this point, the mixed Eulerian-Lagrangian formulation is retained, in which the Eulerian fixed grid describes the flow and the Lagrangian grid (which can be fixed or not) describes the immersed body. These meshes are geometrically independent from each other, and this fact enables to study the flows around simple, complex, movable and deformable geometries, without any remeshing process. These two formulations are coupled by a force field obtained at the Lagrangian points and then distributed over the Eulerian nodes in the body neighborhood. By this strategy, one can use a simple Cartesian grid and it is not necessary to move the grids. According to the VPM [15] the Lagrangian force field is calculated based on the momentum balance over a fluid particle placed on the Lagrangian points.

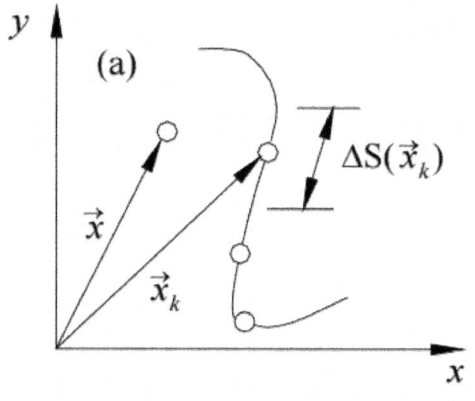

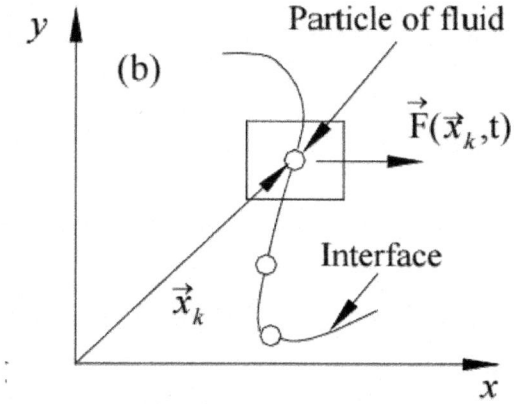

Figure 1: (a) Illustration of the distance between two Lag- rangian points; (b) particle of fluid on the interface.

By considering a particle of fluid placed on the fluid-solid interface as illustrated in Figure 1(b), the Lagrangian force field is given as follows:

$$F(x_k,t) = \rho \underbrace{\frac{\partial V(x_k,t)}{\partial t}}_{F_a} + \underbrace{\rho \nabla [V(x_k,t)V(x_k,t)]}_{F_i} - \underbrace{\mu \nabla [\nabla V(x_k,t) + \nabla^T V(x_k,t)]}_{F_v} + \underbrace{\nabla p(x_k,t)}_{F_p} \quad (4)$$

where F_a [N] is the acceleration force, F_i [N] is the inertial force, F_v [N] is the viscous force, and F_p [N] is the pressure force. After obtaining the force field given by Equation (4), its values are distributed over the Eulerian nodes by using Equation (3) to generate the Eulerian force field that models the immersed body.

NUMERICAL METHOD

A number of mesh-free methods have been developed in recent years [17] to circumvent the polygonisation problem found in the classical numerical methods. Here, the momentum and continuity equations are numerically solved using the finite difference method through the fractional step scheme based on the pressure correction concept [18] . Given the initial velocity, the pressure and the force fields, an estimated velocity field is obtained. This velocity field is used to calculate the pressure correction, by solving a linear system of algebraic equations, for which the Modified Strongly Implicit Procedure (MSI) [19] is used. The Poisson equation gives the coupling between Equation (1) and Equation (2). Also, it provides values of pressure that allow that the velocities components, obtained by using the Navier-Stokes equations, satisfy the mass conservation condition. The time discretization is done by the second order Runge-Kutta method [20] . The estimation of the velocity is calculated as:

$$\frac{\tilde{u}_i^{n+1} - u_i^n}{\Delta t} = -\frac{1}{\rho}\frac{\partial p^n}{\partial x_i} - \left[\frac{\partial(u_i u_j)}{\partial x_j}\right]^n + \frac{\partial}{\partial x_j}\left[\nu\left(\frac{\partial u_i}{\partial x_j} + \frac{\partial u_j}{\partial x_i}\right)\right]^n + f_i^n$$

(5)

where $\tilde{u}_i$ [m/s] is the estimated velocity component, Δt [s] is the computational time step and n is the substep index. The Poisson equation for pressure correction, p'^{n+1}, with the source term given by the divergent of the estimated velocity, is expressed by:

$$\nabla^2 p'^{n+1} = \frac{\rho \nabla \tilde{u}^{n+1}}{\Delta t}$$

(6)

and the velocity field is updated by solving the algebraic equation $u_i^{n+1} = \tilde{u}_i^{n+1} - (\Delta t/\rho)(\partial p'^{n+1}/\partial x_i)$. The previous pressure field p^n and the correction pressure p'^{n+1} are used to calculate the updated values of the pressure field, according to the expression $p^{n+1} = p^n + p'^{n+1}$.

RESULTS AND DISCUSSIONS

In this section, numerical simulations are performed to investigate the effects of the oscillating amplitude and forcing frequency on the flow structure of a circular cylinder. Figure 2(a) shows the computational domain which dimensions are 40d (Lu = 16.5d; Ld = 23.5d), in the streamwise and 15d (H) in the cross-stream direction, where d [m] is the cylinder diameter. The upper and lower boundaries are placed at 7.5d. The flow direction is from the left to the right side of the domain and at the inlet, a uniform velocity profile $(u = 1, v = 0)$ is imposed. A Neumann boundary condition is used at the outlet and lateral boundaries, ($\partial u/\partial x = \partial v/\partial x = 0$, $\partial u/\partial y = \partial v/\partial y = 0$), respectively. For the pressure, the Neumann condition is used at the inlet $(\partial p/\partial x = 0)$ and the Dirichlet condition is used on the outlet and in the lateral boundaries $(p = 0)$. On the surface of the cylinder, no-slip boundary condition $(u = v = 0)$ is virtually employed using the IBM method. The time spent for each numerical simulation was about one hour. Also, the numerical code used herein was developed in Visual C.

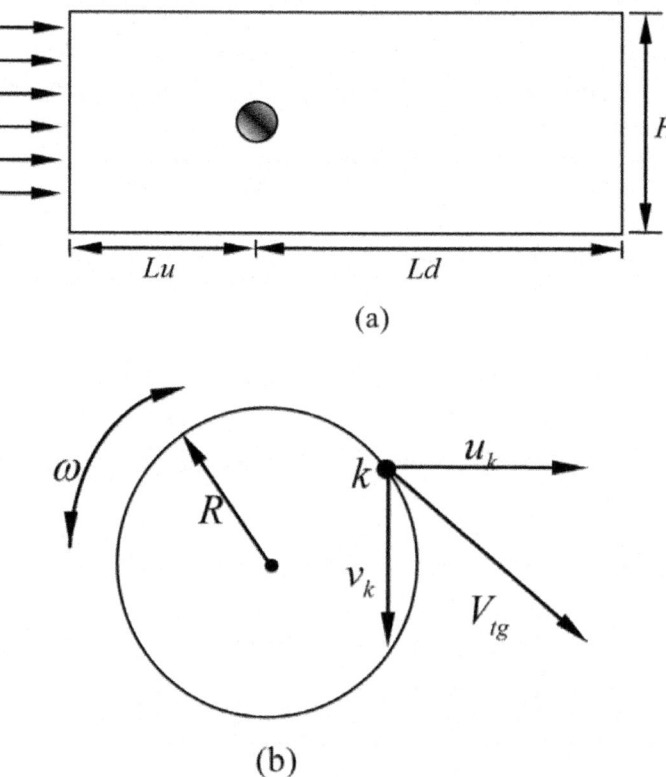

(a)

(b)

Figure 2: (a) Computational domain with a single rotating-oscillating cylinder; (b) Illustration of the velocities components and the angular velocity.

Rotational oscillations at a prescribed set of frequency ratios and amplitudes are then imposed on the cylinder, where the tangential velocity over each Lagrangian point k, as shown in Figure 2(b), is defined as $V_{tg} = \omega R = A\sin(2\pi f_c t)R$, where A [m] is the oscillating amplitude, f_c [Hz] is the forcing frequency, and t [s] is the physical time.

The simulations are performed at Reynolds number 1000, with a time step chosen arbitrarily of $\Delta t = 1 \times 10^{-5} s$ for the first iteration. After that, and during the first 100 iterations, the time step size is increased gradually until $\Delta t = 1 \times 10^{-3} s$, for which the geometry of the immersed cylinder is completely defined according to the immersed boundary methodology. The grid is composed by 400×125 points in x and y directions, respectively; the oscillating amplitude, a, ranges from 1 to 3, while the frequency ratio, $f_r = f_c / f_o$, (f_o is the stationary cylinder frequency) varies from 0 to 6. Many cases have been performed, however, for the purposes of this work, only selected cases are presented.

Vortex Shedding Modes

Figure 3 shows the vorticity contours for $A = 1$ for different frequency ratios. When $f_r = 0$ (Figure 3(a)), it is seen that the vortex street is aligned and symmetric with respect to the central axis of the flow, showing two single vortices shedding per cycle. It is so called 2S mode of vortex shedding. In the frequency range $0.2 \leq f_r \leq 0.6$ (Figure 3(b)) to (Figure 3(d)) the patterns of vortex shedding seem to be similar, but the sizes appear uneven. When $f_r = 0.9$ (Figure 3(e)), one observes that the process of vortex formation is completely different from those of previous cases, and at the distance 12.5d downstream of the cylinder, the 2S mode appears again. Nevertheless, the transversal spacing is greater than that of the stationary case shown in Figure 3(a). When $f_r = 1.05$ (Figure 3(f)), the 2S mode reappear. Figure 3(g) shows the P + S mode in which one pair and a single vortex are shed in each cycle. As frequency ratio further increases, Figure 3(h) shows that the wake structure remains essentially unchanged and similar to the stationary case. The no synchronized flow with the cylinder movement resembles to stationary cylinder flow, with some additional instabilities, due to the cylinder movement [21]. In other words, the instability caused by cylinder oscillation is limited to a region near to the cylinder, while far from immersed body, the vortices redirect to form the stable Von Kármán Street. This implies that occurs vortex-vortex interaction of the same sign, near to the cylinder, resulting in large scale vortices, whose values frequencies are close to the stationary cylinder frequency ($f_o = 0.23$).

Figure 4 shows visualizations of the flow structure obtained for A = 2 in the same frequency range as shown at Figure 3. In the range $0.2 \leq f_r \leq 0.5$, one can note the presence of two vortex pairs shed per cycle, named as 2P mode. When $f_r = 0.6$, the wake vortices are similar to the classical Von Kármán Street shown in Figure 3(a), named 2S mode. However, the longitudinal and transversal spaces are larger than those corresponding to the stationary condition. In the frequency range $0.9 \leq f_r \leq 1.2$, the evolution of the patterns remains unchanged, as shown from Figure 4(d) to Figure 4(f). In addition, it is observed that the transversal spacing is greater near the cylinder and decreases away from it. When $f_r = 2.5$ (Figure 4(g)) it can be noted the presence of two vortex pairs in each side of the central line of the flow, resulting in a conical wake structure. When f_r is higher, it is observed that the vortex pattern is the same as the Kármán vortex street, indicating that for higher frequency values, the effect of rotational oscillation is confined to the flow near the cylinder, and its influence in the far field of the vortex structure is insignificant, as shown in Figure 4(h).

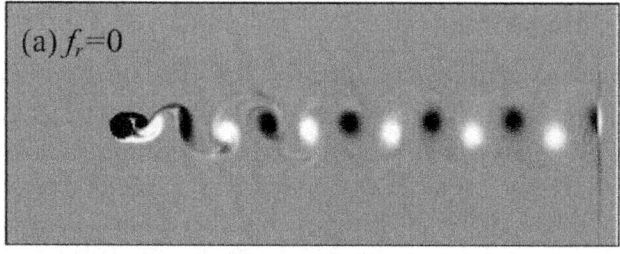

(a) f_r=0

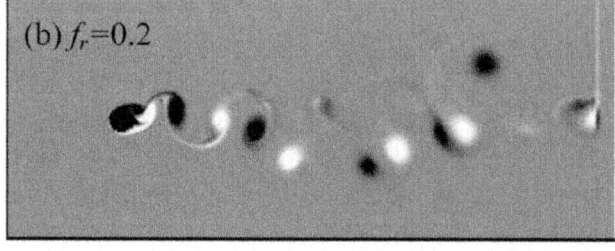

(b) f_r=0.2

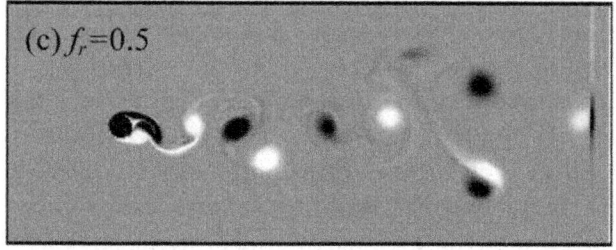

(c) f_r=0.5

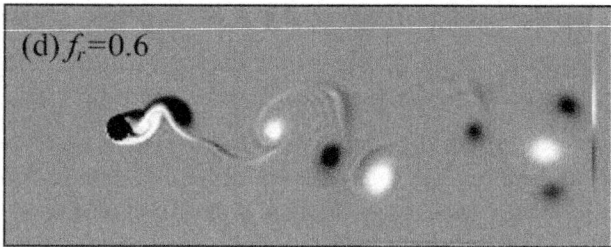

(d) f_r=0.6

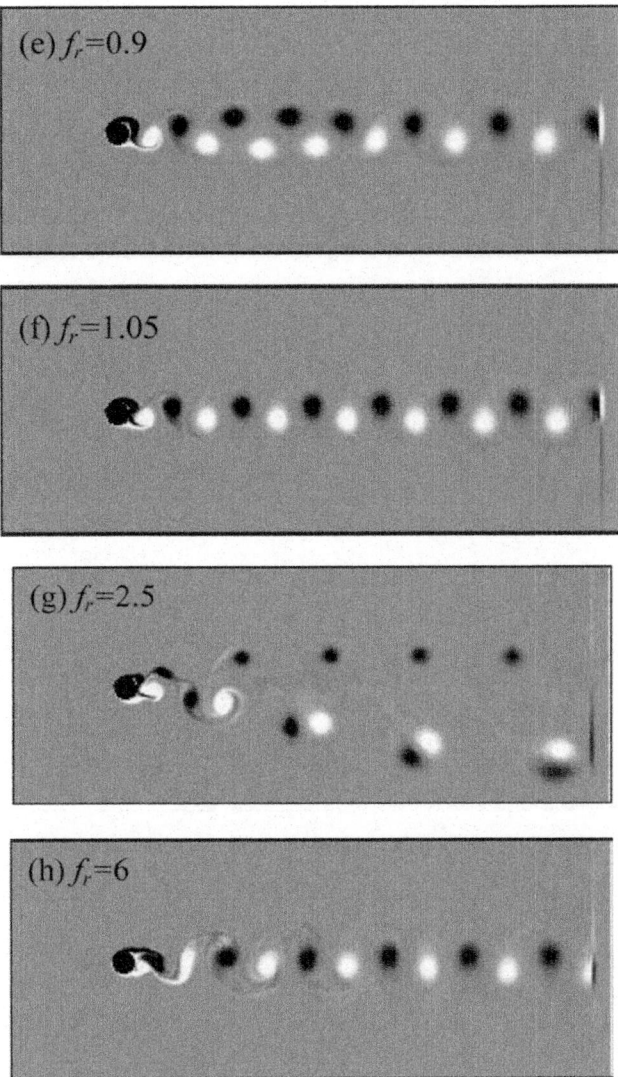

Figure 3. Vorticity contours for Re = 1000 and A = 1 and several values of f_r.

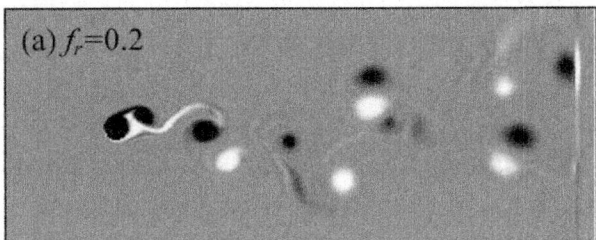

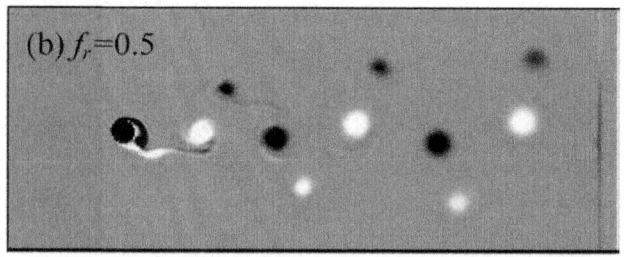

(b) f_r=0.5

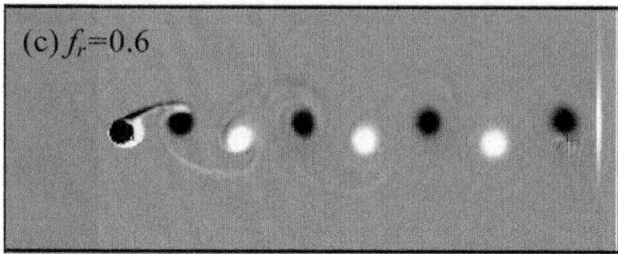

(c) f_r=0.6

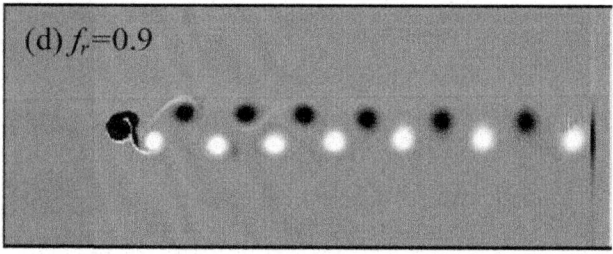

(d) f_r=0.9

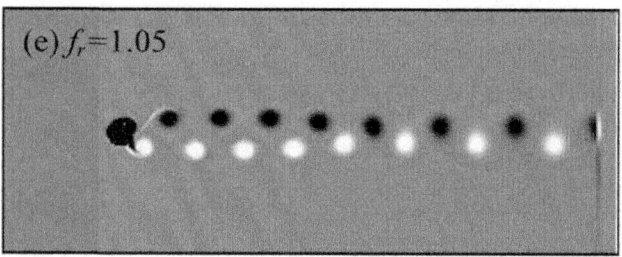

(e) f_r=1.05

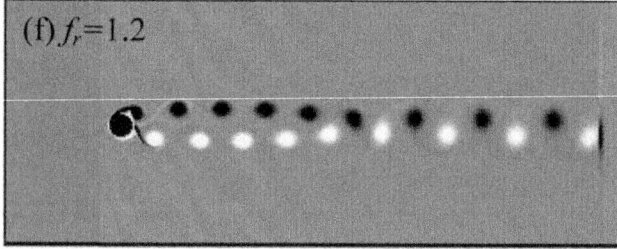

(f) f_r=1.2

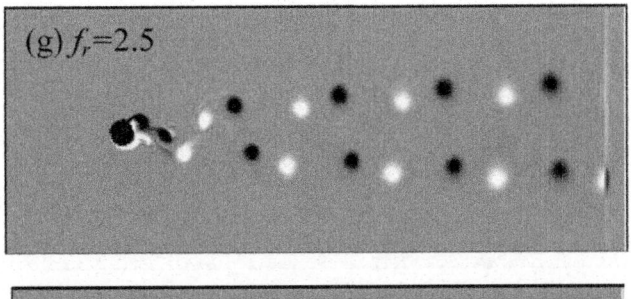

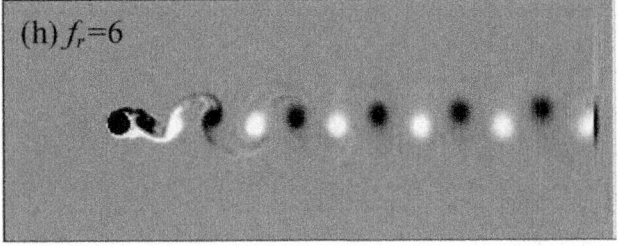

Figure 4. Vorticity contours for Re = 1000 and A = 2 and several values of f_r.

The patterns of vortex shedding from the cylinder in the near and far wakes are shown in Figure 5, for A = 3 and various values of f_r. In Figure 5(a) one can note that the vortex street is not aligned with respect to the central axis of the flow and there are few vortex pairs in the wake. In addition, Figure 5(b) shows that the vortex formation process is completely different from that of the previous cases, in which a new mode of vortex shedding, named as 2 C mode, appears. This mode indicates that two vortex pairs of the same signal are shed per cycle. It is important to mention that this mode was not observed by the authors cited in the references for the case of circular cylinders subjected to rotational oscillations. However, it is noted in the case of a pivoted cylinder as reported in Williamson and Jauvtis [22] .

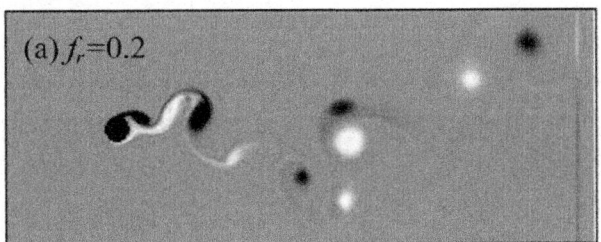

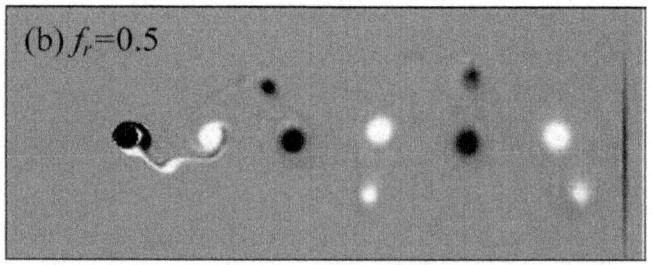

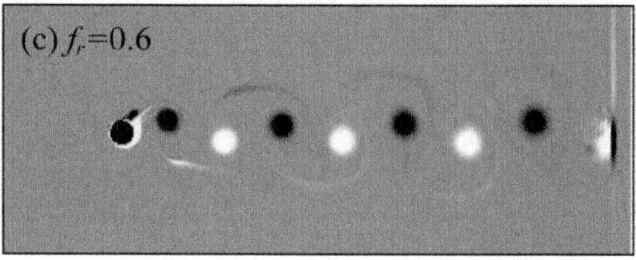

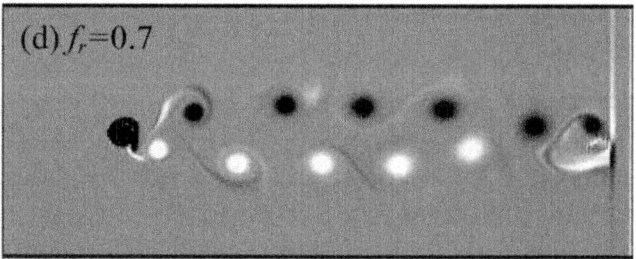

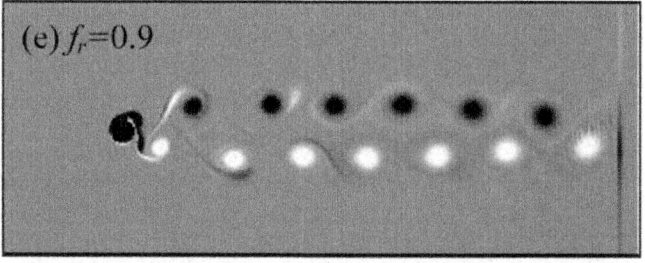

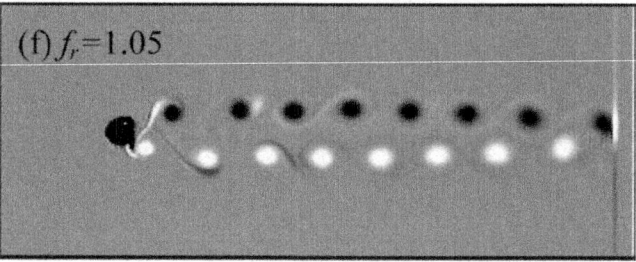

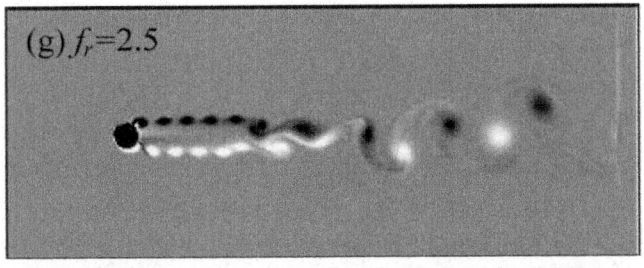

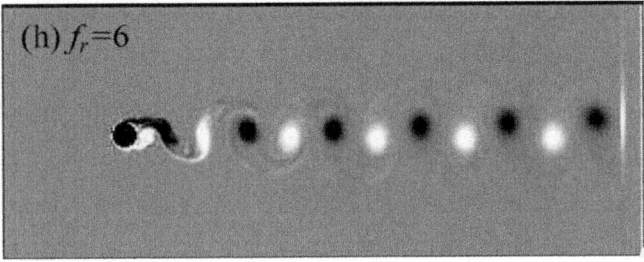

Figure 5: Vorticity contours for Re = 1000 and A = 3 and several values of f_r.

In the frequency range $0.6 \leq f_r \leq 1.05$, one can note that the vortex wakes are found to be basically the same as those obtained for A = 2 in the same frequency range. The result shown inFigure 5(g) demonstrates a new pattern of vortex shedding for which the double wake near the cylinder, composed by vortex of the same signal in each row coalesce at the final double street to form a single wake. As frequency ratio further increases, the pattern of vortex shedding tends to be the same from the non-oscillating case.

Time Histories of the Fluid Dynamics Coefficients and Power Spectra Density

It is known that the vortex shedding process causes fluctuations in the dynamic coefficients and affects the behavior of the flow structure. Figure 6 presents the time histories of the lift (C_l) and drag (C_d) coefficients as well as the power spectra of the lift coefficients for A = 1. Figure 6(a) shows that the transient behavior takes approximately 40 dimensionless time units (nine cycles in the present case) before reaching the periodic regime. From Figure 6(b) to Figure 6(h) one can note that the steady regime is reached more quickly than in the previous case due to fact that the cylinder's oscillations accelerate the vortex shedding process. As can be seen in Figure 6(b), there is no harmonic behavior as observed for the previous stationary case, and the lift curve clearly resembles the shape of a signal with frequency f_e that is beaten by another signal at f_o, indicating that there is a significant interaction between these two frequencies.

Also, from Figure 6(b) to Figure 6(d) it is verified that the amplitudes of the lift coefficient are greater than that for the stationary case. Moreover, it is observed a transition between the two vortex shedding modes when the frequency ratio is increased from $f_r = 1.05$ (Figure 6(f)) (2S mode) to $f_r = 2.5$ (Figure 6(g)) (P + S mode), but the oscillation amplitudes for the P + S mode are larger than that those corresponding to the 2S mode. This means that the change on the vortex modes has a strong influence on the time histories of the dynamics coefficients. As frequency further increases, the oscillation amplitudes tend to be the same of the stationary case, as can be noted by the comparison between Figure 6(h) and Figure 6(a).

When $f_r = 0$, Figure 6(a) shows that the lift spectrum for the stationary case is composed by one peak at the dimensionless frequency $St_o = 0.23$. When $f_r = 0.2$ there are two prominent frequencies in the lift spectrum shown in Figure 6(b). They correspond to the vortex shedding frequency $St_2 = 0.22$, which is closed to St_o and the forcing frequency St_1. As the frequency ratio further increases from $f_r = 0.2$ to $f_r = 0.5$ (Figure 6(c)) shows two frequencies in the power spectrum, but there is a discernible drift of the vortex shedding frequency, St_2, towards the forcing frequency, St_1. The energy level of St_2 (secondary peak) is reduced and the frequency value is 83% of $St_o = 0.23$.

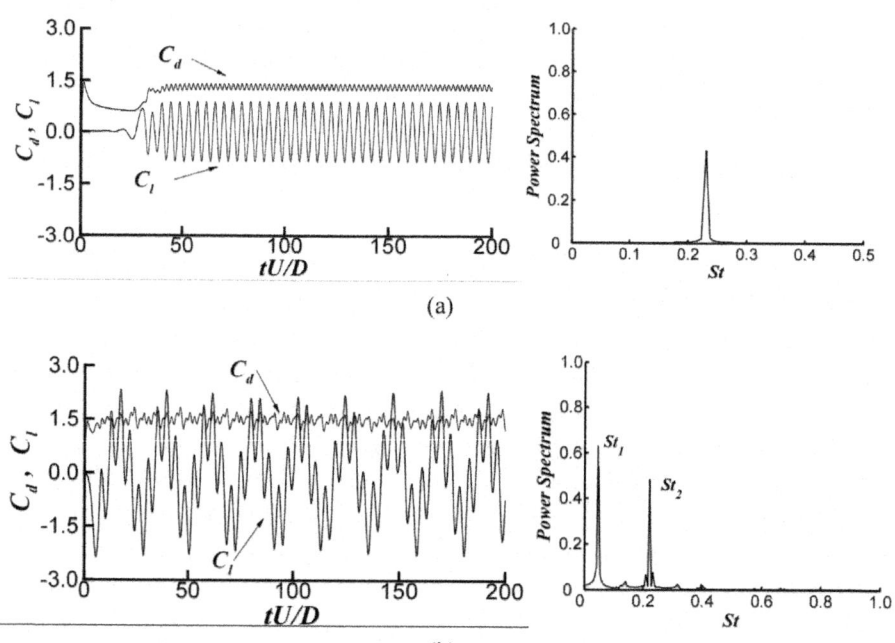

(a)

(b)

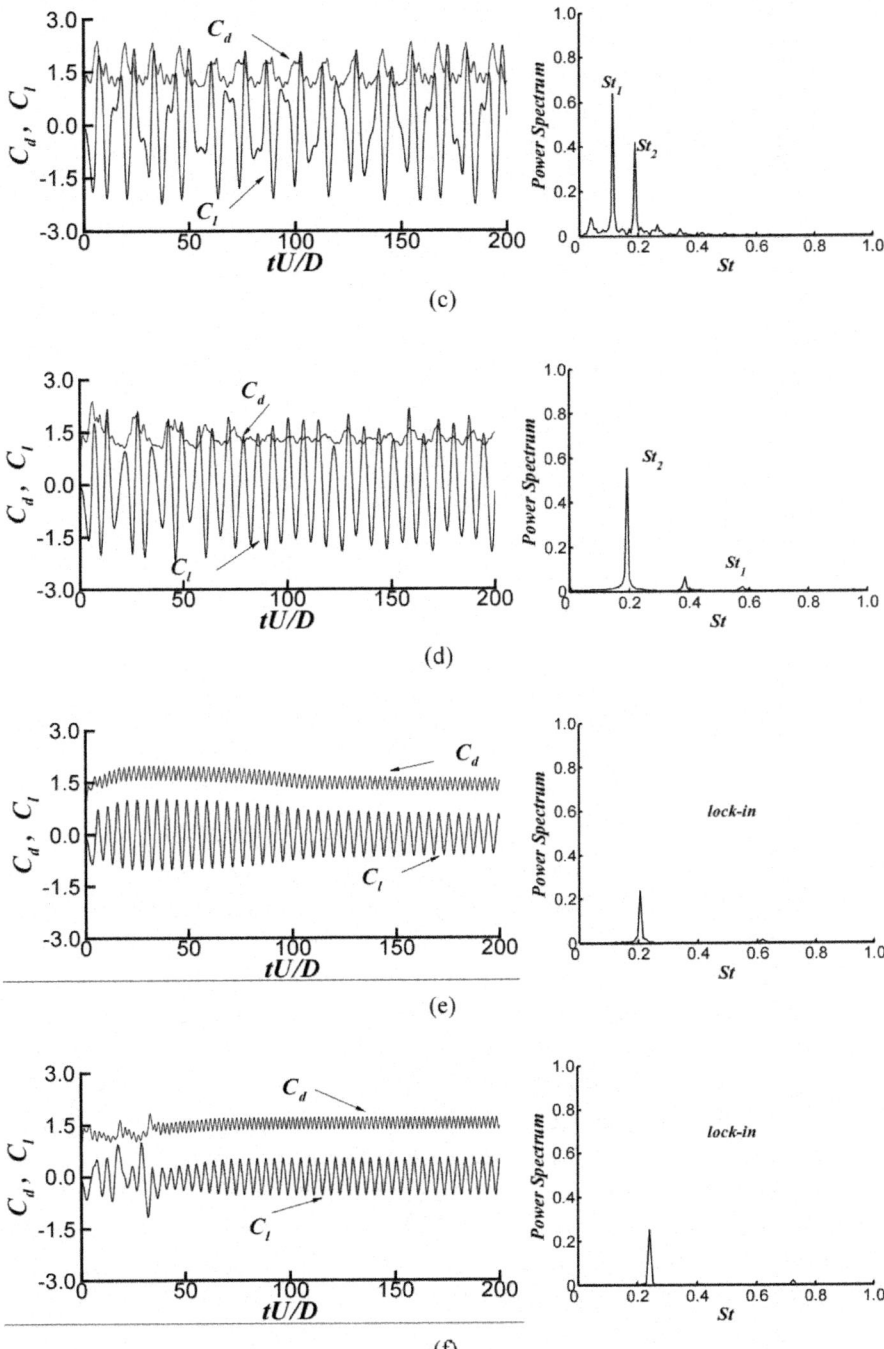

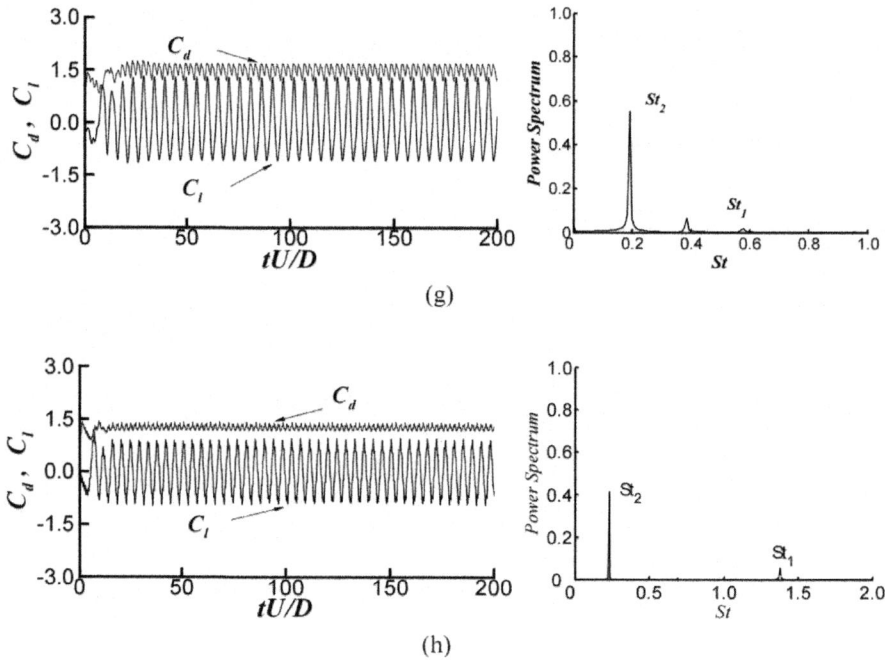

Figure 6. Time histories of the dynamic coefficients and power spectra for A = 1: (a) $f_r = 0$; (b) $f_r = 0.2$; (c) $f_r = 0.5$; (d) $f_r = 0.6$; (e) $f_r = 0.9$; (f) $f_r = 1.05$; (g) $f_r = 2.5$; and (h) $f_r = 6.0$.

When $0.6 \le f_r \le 1.05$, the frequency of vortex shedding, St_2, locks-in to the forcing frequency, St_1, resulting in only one dominating component in the lift frequency spectrum, as shown fromFigure 6(d) to Figure 6(f). During the lock-in regime the wake structure is synchronized with the oscillatory motion of the cylinder and the interaction becomes strong. Also, it is observed that as the frequency ratio increases, the amplitudes of lift coefficient curve decrease and then increase again outside the lock-in regime. Figure 6(d) shows an increase in energy level, compared with the previous, $f_r = 0.5$, followed by a reduction when the frequency ratio is increased, as shown inFigure 6(e) and Figure 6(f). Consequently, this increase and decrease in the frequency peak also affect the mean drag coefficient that reaches a maximum value at $f_r = 0.8$ ($C_d = 1.97$). As frequency ratio further increases, the lock-in regime no longer exists as shown in Figure 6(g) andFigure 6(h). Also, it is interesting to note that when $f_r = 2.5$ (corresponding to the P + S mode), the smallest magnitude of the energy peak corresponds to St_1. On the other hand, as f_r further increases, the magnitude of St_2 tends to be constant, indicating that the large-scale Kármán vortex street has reached some stable state, corresponding to stationary condition.

It is worth mentioning that St_2 is equal to the stationary case $St_o = 0.23$, as shown in Figure 6(h).

Figure 7 shows the time evolution of the dynamics coefficients as well as the power spectra density of the lift coefficient for $A = 2$. When $f_r = 0.2$, the drag and lift coefficients present behaviors similar to those shown in Figure 6(b). Given that each peak (both positive and negative) in the lift curve is related to the shedding of one vortex, then Figure 7(a) means that there should be eighteen vortices in the wake when the dimensionless time reaches 200. By regarding the lift curve, it is also interesting to observe a reduction in the interaction between the two frequencies, from the comparison between Figure 7(a) and Figure 6(b). In the range $0.9 \le f_r \le 1.2$, from Figure 7(d) to Figure 7(f) it is verified that the amplitudes decrease when compared with the previous frequency ratio. When $f_r = 2.5$, the fluctuation of the lift coefficient increases again due to the change of the vortex shedding mode. For higher frequency ratios, the behavior is similar to that observed to $A = 1$, in such a way that the flow tends to the same vortex shedding mode as the one observed for the stationary cylinder. As oscillation amplitude increases, the lock-in regime increases ($0.5 \le f_r \le 1.2$) as shown in Figure 7. The amplitude peak before the lock-in regime for St_1 ($f_r = 0.2$), are larger than those observed in Figure 6(b) and Figure 6(c) for $A = 1$. As f_r increases from $f_r = 0.5$ to $f_r = 1.2$, it is observed a reduction in the magnitude of the peak, as shown in Figure 7(b) to Figure 7(f). During the lock-in regime, the fluctuation amplitude of lift coefficient decreases, as the frequency ratio increases, and increases again outside the lock-in range. It is interesting to point out that in the lock-in regime, for the cases for which has occurred reduction in the magnitude of the energy peak, the wake structure has greater transversal space near the cylinder, while, far away from the cylinder, it decreases (see Figure 4(d) to Figure 4(f)). When $f_r = 2.5$, two spectral peaks can be seen again in the lift spectrum, indicating that the lock-in regime no longer exists. The energy peak associated to St_2 is greater than St_1 and its frequency value ($St_2 = 0.19$) is smaller than that of the non-oscillating case. As f_r further increases, it can be seen in Figure 7(h) that the wake configuration tends to be the same of classical Von Kármán Street corresponding to $St_2 = 0.23$.

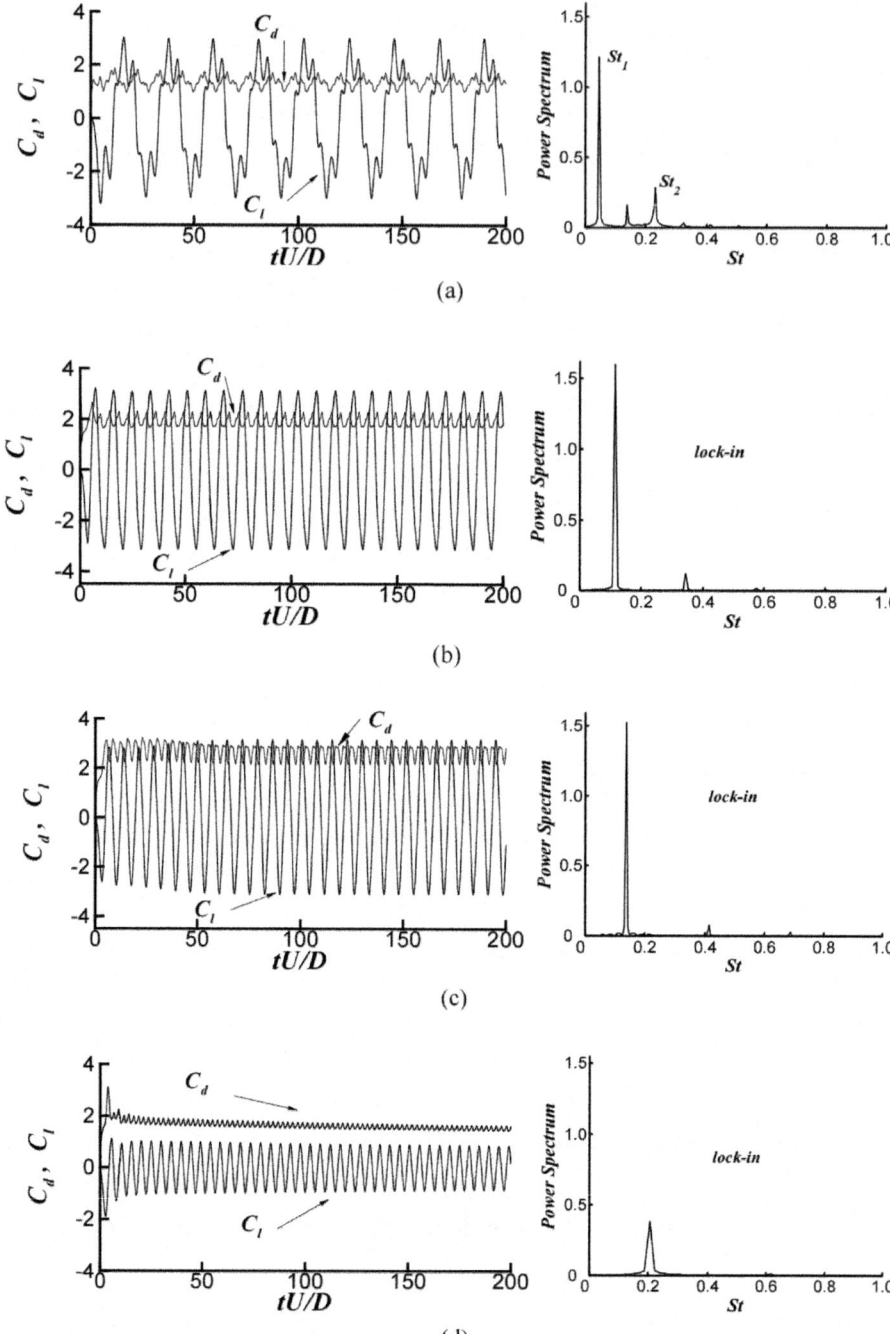

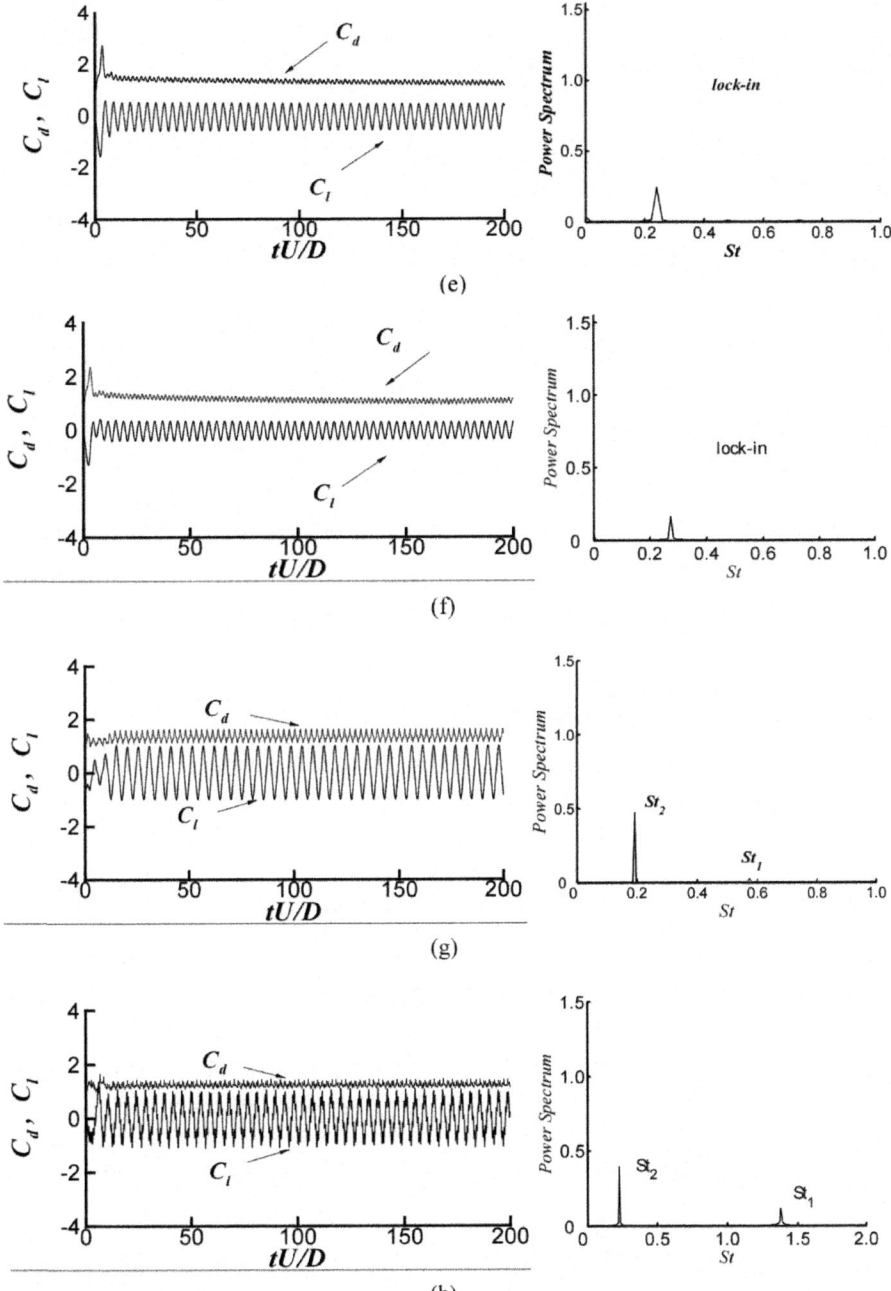

Figure 7: Time histories of the dynamic coefficients and power spectra for A = 2: (a) $f_r = 0.2$; (b) $f_r = 0.5$; (c) $f_r = 0.6$; (d) $f_r = 0.9$; (e) $f_r = 1.05$; (f) $f_r = 1.2$; (g) $f_r = 2.5$; and (h) $f_r = 6.0$.

Figure 8 shows the time histories of the lift and drag coefficients and the power spectra of the lift coefficient for $A=3$. When $f_r = 0.2$, still there are "kink" on the lift curve. This kink phenomenon occurs due to the fact that as the oscillation amplitude increases, a main vortex is formed and shed on one side of the cylinder at the same time as an adjacent secondary vortex having opposite sign is formed and annihilated later by the main vortex. When $f_r = 0.6$, it can be observed a reduction in the amplitude of the lift curve when compared to the previous frequency ratios, $f_r = 0.2$ and $f_r = 0.5$. In the frequency range $0.6 \leq f_r \leq 2.5$ (Figure 8(c) to Figure 8(g)) it is observed that as the frequency ratio increases, the fluctuation amplitude of the lift coefficient is reduced. This reduction is due to the vortex wake structure as shown from Figure 5(c) to Figure 5(g), where it was verified the 2S mode for $f_r = 0.6$, the elliptical wake for $0.7 \leq f_r \leq 1.05$, and a wake with two rows of vortices near the cylinder for $f_r = 2.5$. At higher frequency ratios, the amplitude of the lift curves approaches to that observed for the stationary case.

For $A=3$, the lock-in regime is located at the interval $0.2 \leq f_r \leq 2.5$, for which the maximum mean drag coefficient is obtained at $f_r = 0.6$ ($C_d = 2.93$). Also, as the frequency ratio increases from $f_r = 0.2$ to $f_r = 0.5$, the energy level increases. After that, the energy level reduces from $f_r = 0.5$ to $f_r = 2.5$. Finally, another aspect to be pointed out is that as the oscillation amplitude increases, the frequency range in which the lock-in regime occurs is enlarged: $[0.6-1.05]$ for $A=1$; $[0.5-1.2]$ for $A=2$ and $[0.2-2.5]$ for $A=3$.

Another relevant aspect to be investigated is the energy level of the power spectra, as shown inFigure 9. The most immediate conclusion is that as the frequency ratio increases, the energy level increases and after decreases keeping approximately unchanged for all analyzed amplitudes, for frequency ratio equal to 2. Also, higher energy levels are obtained in the lock-in regime for all amplitudes. Moreover, as the oscillating amplitude increase the energy level in the lower boundary of the lock-in regime also increase.

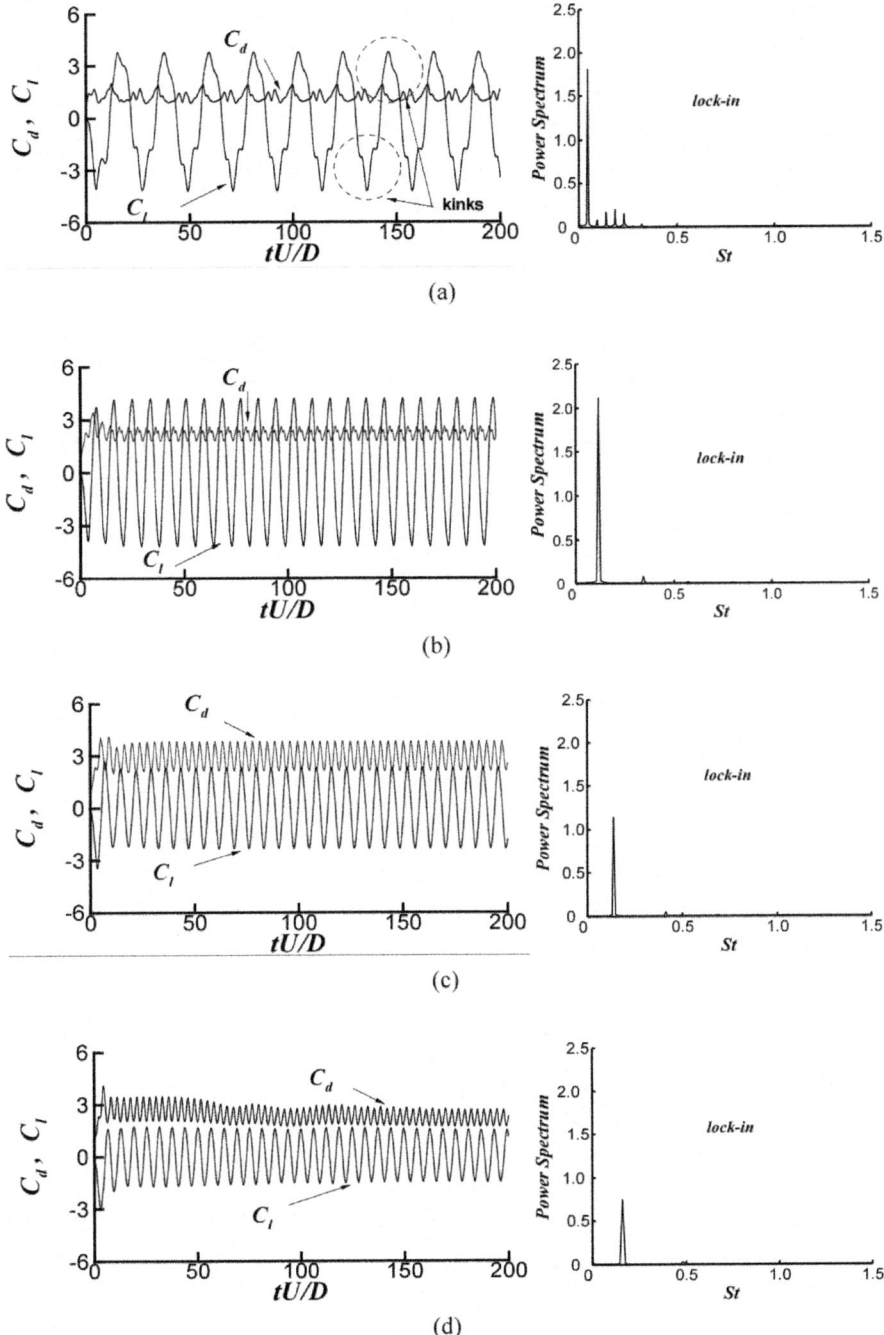

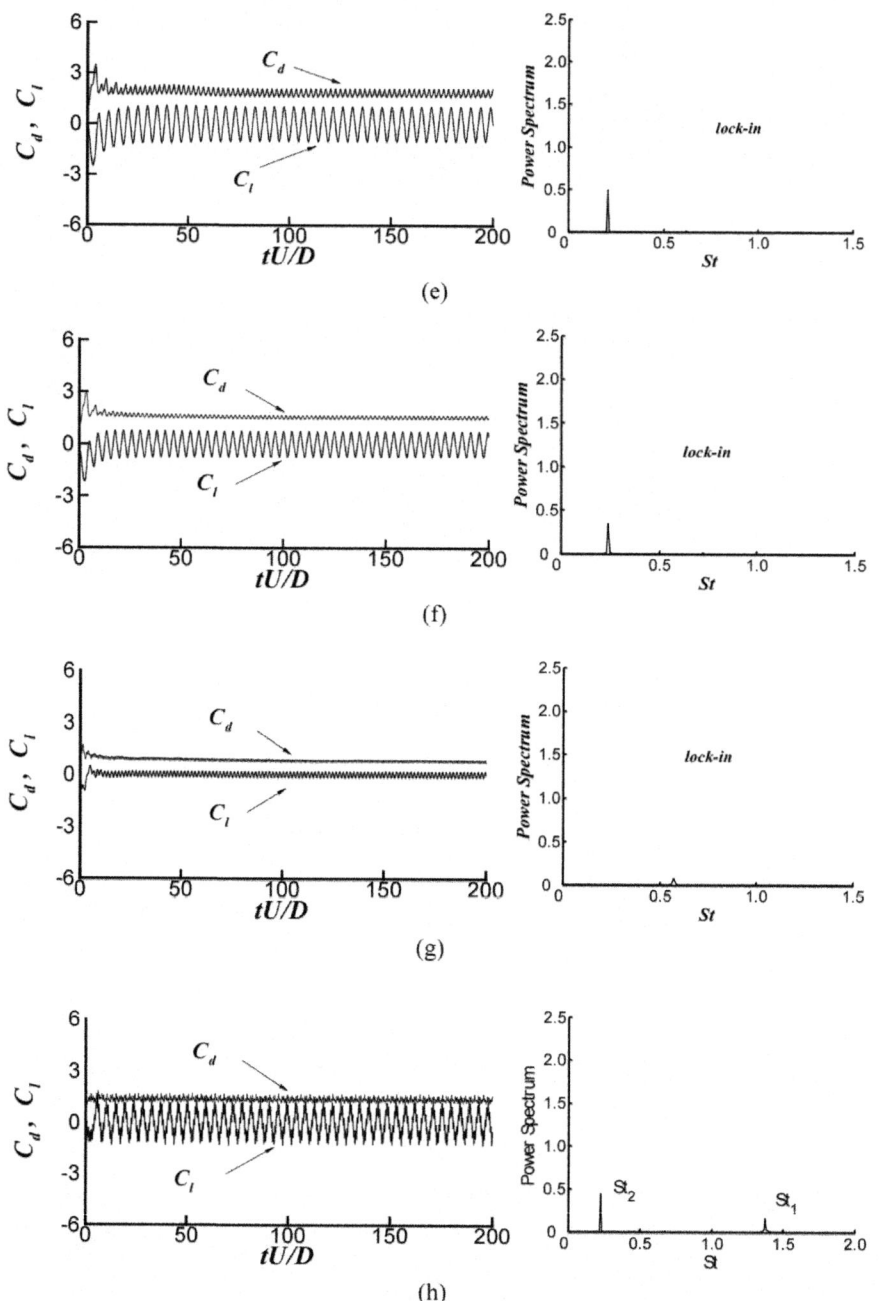

Figure 8: Time histories of the dynamic coefficients and power spectra for $A = 3$: (a) $f_r = 0.2$; (b) $f_r = 0.5$; (c) $f_r = 0.6$; (d) $f_r = 0.7$; (e) $f_r = 0.9$; (f) $f_r = 1.05$; (g) $f_r = 2.5$; and (h) $f_r = 6.0$.

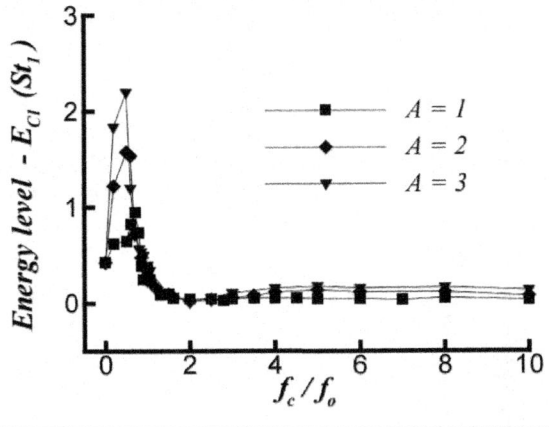

Figure 9: Energy level of St_1 in function of frequency ratio fr for A = 1, A = 2 and A = 3.

Comparison with other Previous Studies

It is desirable to compare the present results with some numerical results from previous studies reported by other investigators. Figure 10 (left) shows a plot of the mean drag coefficient as a function of the frequency ratio obtained for the present numerical simulations and the literature results [4] [9] , for $A = 3$ at Reynolds number 1000. It can be noted that all the results indicate a sharp peak of the C_d curve at low frequency ratios, and small values of C_d at high frequencies ratio This behavior has been also observed for other Reynolds number and oscillation amplitudes [10] [23] . It can be verified that the present computational results are in good agreement with the results by references for low frequency ratios, but overestimates C_d in high frequencies, which may reflect the assumptions of the numerical methodology used in the present calculations. It is important to mention that according to Srinivas and Fujisawa [10] , there is a great discrepancy in the literature concerning the behavior of the mean drag coefficient in the Reynolds numbers range [1000-3000].

Another aspect to be pointed out is that the maximum mean drag coefficients were obtained for the 2S mode for all the analyzed amplitudes, in which the longitudinal and transversal spacing for the oscillating cylinder is greater than those of the stationary cylinder. This fact can be observed by analyzing the vorticity contours corresponding to the maximum C_d, as shown in Figure 11, where a is the transversal spacing and b indicates the longitudinal spacing. It should be noted that as the oscillating amplitude increases, b increases and a takes a constant value. This enables to conclude that the mean drag coefficient is more strongly dependent on the longitudinal spacing than the transversal spacing.

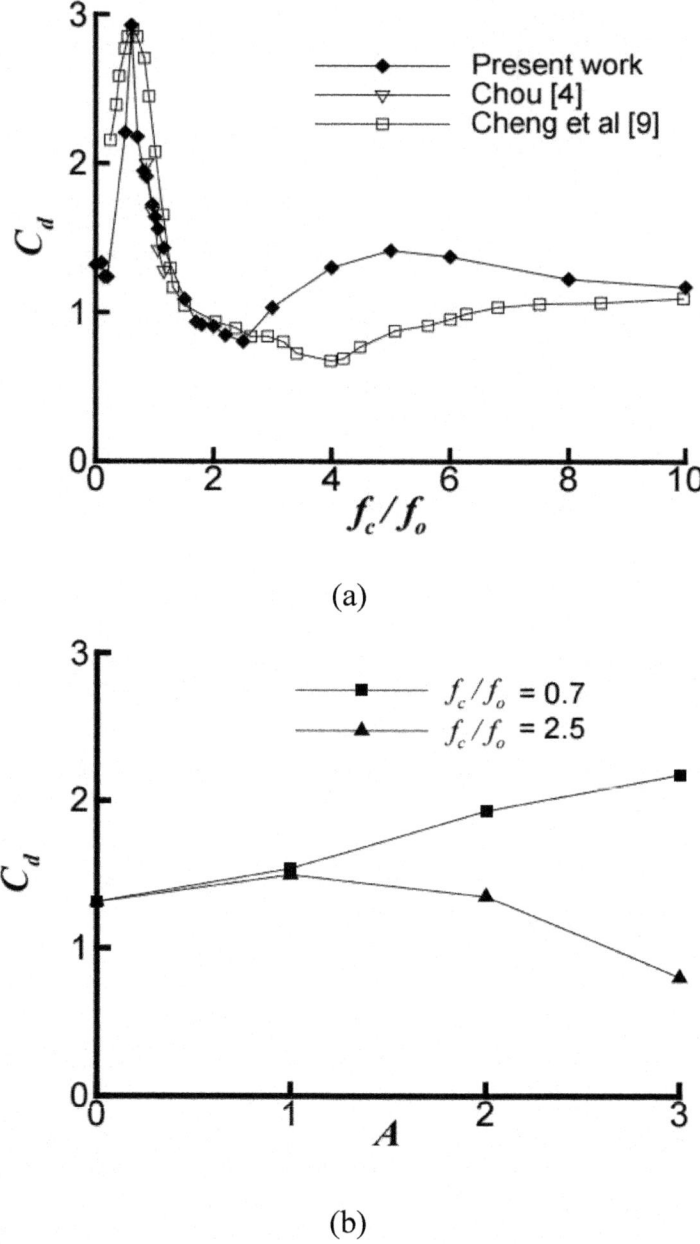

(a)

(b)

Figure 10: (left) Mean drag coefficient in function of f_r for A = 3 and (right) in function of oscillation amplitude. Re = 1000.

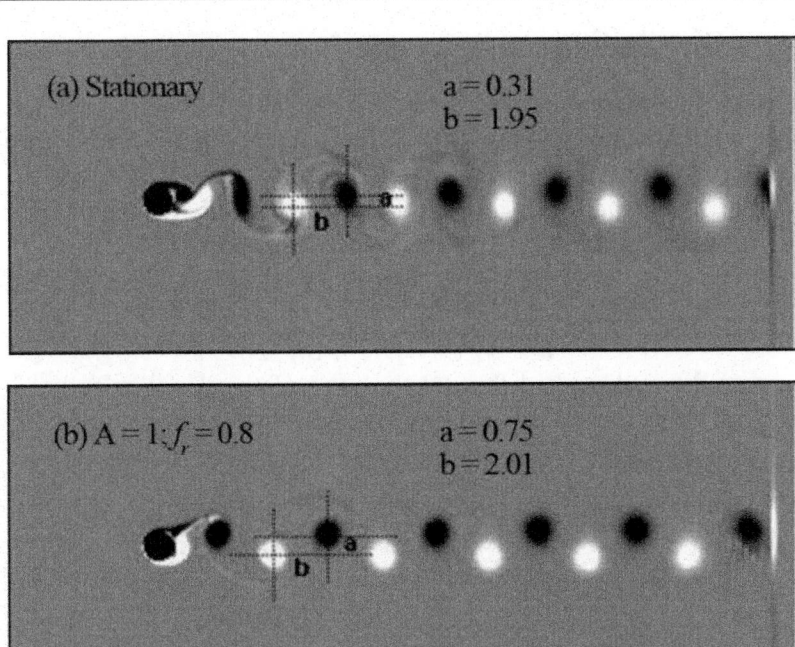

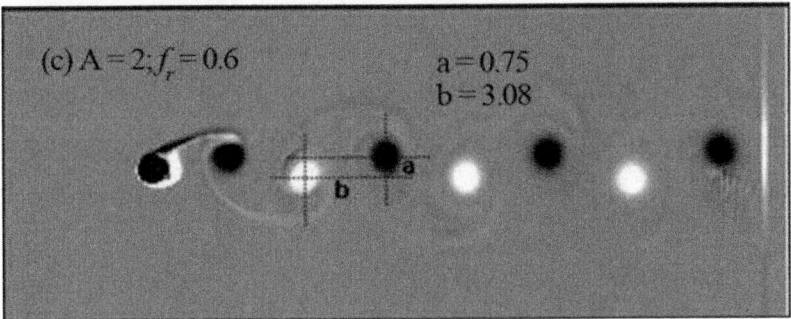

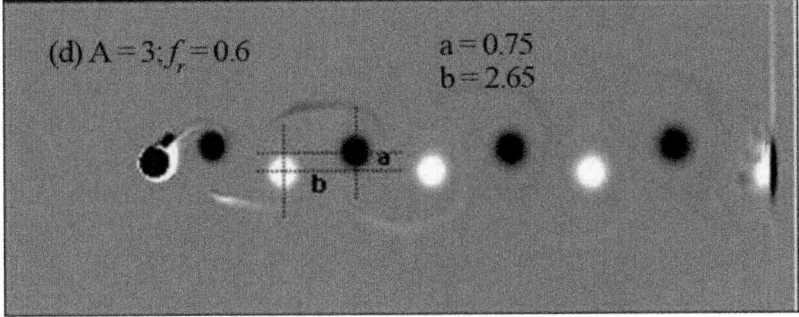

Figure 11: Presentation of longitudinal and transversal spacing between the vortices.

Pressure Distribution

The mechanism of increase and reduction of C_d for low and high frequency ratios, respectively, can also be analyzed through the mean pressure distribution along the surface of the stationary and oscillating cylinders. Figure 12(a) shows an illustrative scheme of the angle θ along the surface of the immersed body, which is defined as zero at the downstream point of the cylinder.

The pressure coefficient is obtained by the relation $C_p = (p_k - p_\infty)/(0.5\rho U_\infty^2)$, where p_k is the static pressure in the Lagrangian points, p_∞ is the static pressure of the undisturbed flow at the inlet of the domain, and $(0.5\rho U_\infty^2)$ is the dynamic pressure taken as the reference. Figure 12(b) and Figure 12(c) show the distribution of the mean pressure coefficient C_p for all the analyzed amplitudes and two frequency ratios. It can be noted that C_p at the maximum local pressure ($\theta = 0°$) is almost unit, as expected. Also, it can be verified how the mean pressure coefficient downstream of the cylinder ($\theta = 180°$) varies with the cylinder oscillation amplitude. When $f_r = 0.6$, it can be observed that for A = 1 the mean pressure coefficient downstream of the cylinder ($\theta = 180°$) presents a small decrease (in absolute values) when compared with the stationary cylinder. Nevertheless, it is noted a considerable increase in the mean pressure coefficient when the oscillation amplitude increases from A = 1 to A = 2. As amplitude further increases from A = 2 to A = 3, C_p tends to be the same for $\theta = 180°$. When $f_r = 1.5$, it can be observed a little decrease in C_p at $\theta = 180°$ for A = 1, while for A = 2, a greater reduction in C_p value is observed from $C_p = -2.84$ ($f_r = 0.6$) to $C_p = -0.25$ ($f_r = 1.5$).

By comparing Figure 12(b) and Figure 12(c) for A = 3, it can be noted an inversion of the mean pressure coefficient signal, from $C_p = -2.8$ ($f_r = 0.6$) to $C_p = 0.04$ ($f_r = 1.5$). Thus, the forcing frequency plays an important role on the pressure distribution, enabling to verify that for low frequency ratios, the flow downstream of the cylinder is very dissipative due to viscous effects. This effect contributes to reduce the pressure and consequently to increase the drag over the cylinder. However, at high frequency ratios, the wake dynamic downstream of the cylinder is not so strong when compared to the corresponding ones obtained for low frequencies, which contribute to increase the pressure and, consequently, to reduce the drag coefficient.

CONCLUDING REMARKS

The numerical simulations of the flow over the rotationally-oscillating circular cylinder by using the Immersed

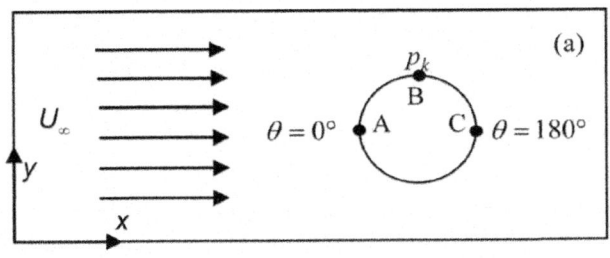

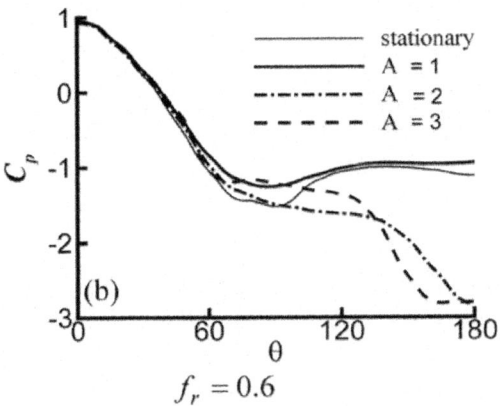

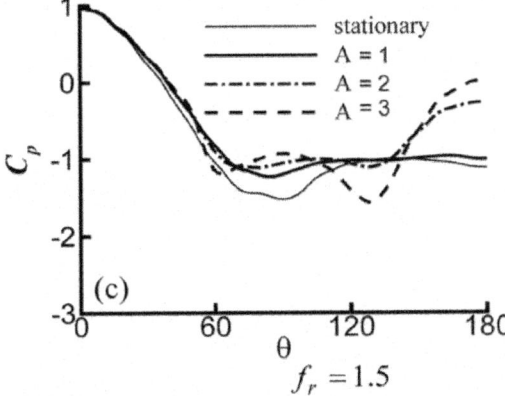

Figure 12: Illustrative scheme of the angle q along the cylinder surface (a) and the distribution of the mean pressure coefficient in function of the angle (b) and (c).

Boundary Method combined with the Virtual Physical Model have been addressed. Given the strong influence of the oscillation amplitude and frequency on the flow around the cylinder, the influence of these parameters on the lift and drag coefficients, on the pressure distribution, as well as on the vortex shedding frequency has been investigated.

The simulations shown that the flow structure in the near wake was strongly dependent on the oscillation frequency. Also, it observed different vortex shedding modes ("2C", "2P", "2S", and "P + S") for a fixed oscillation amplitude at different values of frequency ratios, and also for a fixed frequency ratio at different values of oscillation amplitudes, in addition to the conical wakes. Another important feature of the numerical methodology is its capability to identify the 2 C mode of the vortex shedding, which has not been observed by other investigators for the case of rotationally-oscillating cylinders.

The study reported herein enabled to observe a number of important features that should be mentioned:

- The range of resonance increases as the oscillation amplitude increases;
- For high frequency ratios, the wake structure configuration is similar to the classical Von Kármán Street, and the values of the vortex shedding frequency take the values corresponding to the stationary cylinder case;
- The pressure distribution over the cylinder is influenced by forcing frequency, and consequently affects the drag over the cylinder. It implies that drag control can be done by the rotational oscillations mechanism;
- The Immersed Boundary Method combined with the Virtual Physical Model can be easily employed in the case of moving bodies, being a very useful tool to simulate problems involving prescribed motion.

ACKNOWLEDGEMENTS

The author is grateful to the following organizations: Agency of the Ministry of Science, Technology and Innovation—CNPq for the continued support to their research work, and the Minas Gerais State Agency FAPEMIG.

REFERENCES

1. Anagnostopoulos, P. (2002) Flow-Induced Vibrations in Engineering Practice. WIT Press, Southampton, Boston.

2. Naudascher, E. and Rockwell, D. (1994) Flow-Induced Vibrations: An Engineering Guide. Dover Publications, Inc., Mineola, New York.

3. Païdoussis, M.P. (2004) Fluid-Structure Interactions: Slender Structures and Axial Flow. Vol. 2, Elsevier Academic Press, San Diego.

4. Chou, M.H. (1997) Synchronization of Vortex Shedding from a Cylinder under Rotary Oscillation. Computers & Fluids, 36, 755-774. http://dx.doi.org/10.1016/S0045-7930(97)00028-5

5. He, J.W., Glowinski, R., Metcalfe, R., Nordlander, A. and Periaux, J. (2000) Active Control and Drag Optimization for Flow Past a Circular Cylinder I. Oscillatory Cylinder Rotation. Journal of Computational Physics, 163, 83-117. http://dx.doi.org/10.1006/jcph.2000.6556

6. Lee, S.-J. and Lee, J.-Y. (2006) Flow Structure of Wake behind a Rotationally Oscillating Circular Cylinder. Journal of Fluids and Structures, 22, 1097-1112. http://dx.doi.org/10.1016/j.jfluidstructs.2006.07.008

7. Du, L. and Dalton, C. (2013) LES Calculation for Uniform Flow past Rotationally Oscillating Cylinder. Journal of Fluids and Structures, 42, 40-54. http://dx.doi.org/10.1016/j.jfluidstructs.2013.05.008

8. Cheng, M., Liu, G.R. and Lam, K.Y. (2001) Numerical Simulation of Flow Past a Rotationally Oscillating Cylinder. Computers & Fluids, 30, 365-392. http://dx.doi.org/10.1016/S0045-7930(00)00012-8

9. Cheng, M., Chew, Y.T. and Luo, S.C. (2001) Numerical Investigation of a Rotationally Oscillating Cylinder in Mean Flow. Journal of Fluids and Structures, 15, 981-1007. http://dx.doi.org/10.1006/jfls.2001.0387

10. Srinivas, K. and Fujisawa, N. (2003) Effect of Rotational Oscillation upon Fluid Forces about a Circular Cylinder. Journal of Wind Engineering and Industrial Aerodynamics, 91, 637-652. http://dx.doi.org/10.1016/S0167-6105(02)00460-9

11. Ray, P. and Christofides, P.D. (2005) Control of Flow over a Cylinder Using Rotational Oscillations. Computers and Chemical Engineering, 29, 877-885. http://dx.doi.org/10.1016/j.compchemeng.2004.09.014

12. Peskin, C.S. (1977) Numerical Analysis of Blood Flow in the Heart. Journal of Computational Physics, 25, 220-252. http://dx.doi.org/10.1016/0021-9991(77)90100-0

13. Nicolás, A. and Bermúdez, B. (2007) Viscous Incompressible Flows by the Velocity-Vorticity Navier-Stokes Equations. CMES: Computer Modeling in Engineering & Sciences, 20, 73-83.

14. Báez, E. and Nicolás, A. (2009) Recirculation of Viscous Incompressible Flows in Enclosures. CMES: Computer Modeling in Engineering & Sciences, 41, 107-130.

15. Lima e Silva, A.L.F., Silva, A.R. and Silveira-Neto, A. (2007) Numerical Simulation of Two-Dimensional Complex Flows around Bluff Bodies Using the Immersed Boundary Method. Journal of the Brazilian Society

of Mechanical Sciences and Engineering, XXIX, 378-386. http://dx.doi.
org/10.1590/s1678-58782007000400006

16. Peskin, C.S. and McQueen, D.M. (1994) A General Method for the
    Computer Simulation of Biological Systems Inter-acting with Fluids.
    SEB Symposium on Biological Fluid Dynamics, Leeds, England, 5-8
    July 1994.

17. Vertnik, R. and Sarler, B. (2009) Solution of Incompressible Turbulent
    Flow by a Mesh-Free Method. CMES: Computer Modeling in Engineering
    & Sciences, 44, 65-95.

18. Chorin, A. (1968) Numerical Solution of the Navier-Stokes Equations.
    Mathematics of Computations, 22, 745-762. http://dx.doi.org/10.1090/
    S0025-5718-1968-0242392-2

19. Schneider, G.E. and Zedan, M.A. (1981) Modified Strongly Implicit
    Procedure for the Numerical Solution of Field Problems. Numerical Heat
    Transfer, 4, 1-19. http://dx.doi.org/10.1080/01495728108961775

20. Ferziger, J.H. and Peric, M. (2002) Computational Methods for Fluid
    Dynamics. 3rd Edition, Springer-Verlag, Berlin, 423 p. http://dx.doi.
    org/10.1007/978-3-642-56026-2

21. Tuszynski, J. and Löhner, R. (1998) Control of a Kármán Vortex Flow by
    Rotational Oscillations of a Cylinder. George Mason University, USA,
    1-12.

22. Williamson, C.H.K. and Jauvtis, N. (2004) A High-Amplitude 2T Mode
    of Vortex-Induced Vibration for a Light Body in X-Y Motion. European
    Journal of Mechanics—B/Fluids, 23, 107-114. http://dx.doi.org/10.1016/j.
    euromechflu.2003.09.008

23. Fujisawa, N., Asano, Y., Arakawa, C. and Hashimoto, T. (2005)
    Computational and Experimental Study on Flow around a Rotationally
    Oscillating Circular Cylinder in a Uniform Flow. Journal of Wind
    Engineering and Industrial Aerodynamics, 93, 137-153. http://dx.doi.
    org/10.1016/j.jweia.2004.11.002

Chapter 3

AERODYNAMIC BRAKE FOR FORMULA CARS

Roberto Capata and Leone Martellucci

Department of Mechanical and Aerospace Engineering, University of Roma "Sapienza", Rome, Italy

ABSTRACT

In the last years, in formula racing cars championships, the aerodynamic had reached an ever more important stance as a performance parameter. In the last four seasons, Red Bull Racing Technical Officer had designed their Formula 1 car with the specific aim to generate the optimal downforce, in relation to the car instantaneous setup. However, this extreme research of higher downforce brings some negative effects when a car is within the wake of another car; indeed, it is well known that under these condition the aerodynamic is disturbed, and it makes difficult to overtake the leading car. To partially remedy this problem, Formula 1 regulations introduced the Drag Reduction System (DRS) in 2011, which was an adjustable flap located on the rear wing; if it is flattened, allowing to reduce the downforce, increasing significantly the velocity and, therefore, the chances to overtake the leading car. Vice versa, when the flap is closed, it ensures a higher grip, which is very useful especially in medium-slow speed turns. Keeping the focus on the rear wing, but by shifting attention from the increased top speed to increase the grip in the middle and slow speed curves, we decided to study a similar device to the DRS, but with the opposite effect. The aim is to design an aerodynamic brake integrated with the rear wing. In particular, the project idea was to sculpt, on the upper surface of the wing (pressure side), a series of "C" shaped cavity, normally covered by adequate sliding panels. These cavities, when they are discovered, at the beginning of the braking phase, produce a turbulence and additional increase downforce, lightening the load on the braking system and allowing the pilot to substantially reduce slippage and to delay the braking. Since it seems that the regulations adopted by the FIA Formula 1 Championship do not allow such a device, it has been decided to apply the concept on a Formula 4 vehicle.

This paper describes the design and analyzes the effects of these details on a standard wing cavity, using a commercial CFD software.

PROBLEM FORMULATION

In this paper, the realization of an aerodynamic brake integrated in a rear wing of a formula car has been considered. The first step consists in the choice of an appropriate aerodynamic appendix. In particular, it was decided to study an Italian Formula 4 race car [1] , being a category in the first stages of development. Also, the regulation of this championship is easy to find and the car is characterized by uniformity of the mechanics and the airfoils. Therefore, taken note of the technical regulation on FIA website, it was decided to study the upper airfoil, of which was shown a dimensioned drawing (Figure 1). It is an aluminum alloy wing, with a chord line of 237.9 mm and a height of 54.2 mm.

Formula 4 championship will provide the use of a 4T heat engine (Otto/ Beau de Rochas cycle): it can be naturally aspirated or turbocharged, with maximum power in the order of 120 kW (160 HP). Considering the weight of the car and the race tracks of the championship, it is predicted a maximum speed of 230 km/h (64 m/s). Regarding the operating conditions, an air temperature of 300K was assumed at atmospheric pressure.

Briefing Description of Airfoil Behavior

Considering an airfoil, there are several elements that have a specific nomenclature:

1. Mean camber line: locus of points halfway between the upper and lower surface as measured perpendicular to the mean chamber line itself;
2. Leading edge: the most forward point of the mean camber line;
3. Trailing edge: the rearmost point of the mean camber line;
4. Chord: the straight line joining the leading edge with the trailing edge;
5. Upper surface: the upper boundary of the profile;
6. Lower surface: the lower boundary of the profile;
7. Thickness: the distance between the lower surface and the upper surface.

The different airfoil shapes are marked by a logical numbering system which was introduced by the U.S. federal agency NACA. This system consists of four digits which have a definite meaning:

- the first digit indicates the maximum camber in hundredths of chord;

- the second digit represents the location of maximum camber along the chord from leading edge in tenths of chord;
- the third and fourth give the maximum thickness in hundredths of chord.

When an airfoil is moving relative to the air, it generates an aerodynamic force, in a rearward direction at an angle with the direction of relative motion. This aerodynamic force is commonly resolved into two components: lift and drag. Lift is the force component perpendicular to the direction of relative motion while Drag is the force component parallel to the direction of relative motion. These forces are studied at different angles of attack which is the angle at which an airfoil cleaves fluid. The experimental data show that CL varies with the angle of attack: more precisely, at low angles of attack the lift coefficient CL varies linearly with α. In a region characterized by a linear trend, the flow moves smoothly over the airfoil and is attached to the back of the wing. As soon as α increases, the flow tends to separate from the surface of the airfoil, creating a region of "dead air" behind the profile. A briefing flow analysis of the physical phenomenon in question in order to understand better what is happening in the latter case is reported. It is clear from Figure 2 that the speed at the trailing edge tends to increase, with a strong reduction of the pressure, while in the stagnation point the speed tends to be zero and pressure rises sharply. It creates an adverse pressure gradient, thus particles of fluid move from the trailing edge to the stagnation point, and then it has a rapid separation of the boundary layer below. Stagnation point does not have a stable position in these conditions because there is not pressure recovery. The recirculation generated by the detachment of the boundary layer creates first vortex that causes a wake vortex. It is necessary to study the turbulent behavior of the fluid that meets the wing, through the Navier-Stokes equations in order to consider the stall of the wing:

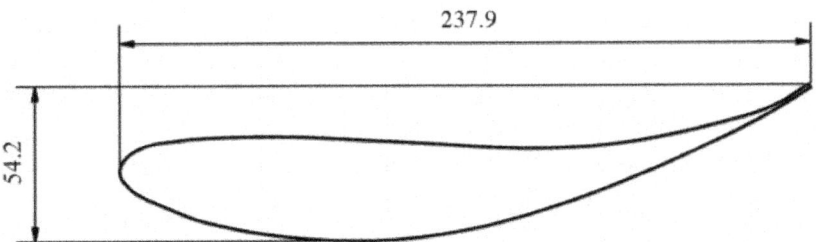

Figure 1: Dimensioned Drawing of a F4 rear wing (in mm).

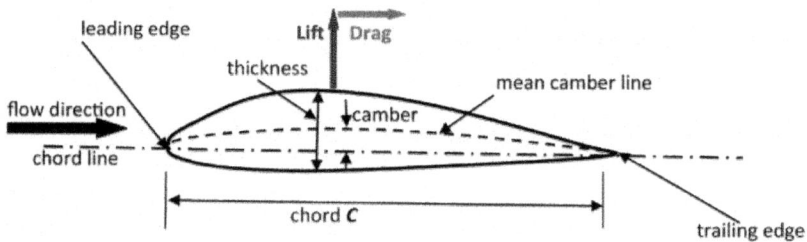

Figure 2: Airfoil characteristics with generic flow direction.

$$\rho \cdot \frac{\partial u_i}{\partial t} + \rho \cdot u_j \cdot \frac{\partial u_i}{\partial t} = \rho \cdot f_i - \frac{\partial p}{\partial x_i} + \mu \cdot \frac{\partial^2 u_i}{\partial x_j \cdot \partial x_j}$$

(1)

where u(x, t) is the instantaneous velocity, ρ the medium density, μ the viscosity and f the applied force.

This system of equations is a system of partial differential equations that describe the behavior of a Stokesian fluid: the fluid can be considered to be continuous. There is an analytical solution only in simplified cases, while solutions in the other cases can be obtained using simplified methods of numerical analysis. The most straight- forward method for the numerical simulation of turbulent flows is direct numerical simulation DNS which discretizes the Navier-Stokes equations. It resolves the entire range of turbulent length scales thus the description of the flow is so detailed that the validity of the simulation is similar to an experiment. The computational cost is proportional to Re^3, thus it is necessary to use a different solution studying turbulent flows at high Reynolds, because the computational resources required by a DNS would exceed the capacity of the most powerful computer currently available. In practical applications, the knowledge of the average quantities is enough to solve the problem of a turbulent flow; the basic idea of the technique RANS (Reynolds Averaged Navier-Stokes Equations) is to derive only the average parameters (mediated in time) from Navier-Stokes equations, reducing the enormous computational cost required by DNS. In practice, the turbulent motion consists of a mean motion and fluctuation over time. Using the decomposition of Reynolds:

$$u(x,t) = \langle u(x,t) \rangle + u'(x,t)$$

(2)

where $u(x,t)$ is the instantaneous velocity, $\langle u(x,t) \rangle$ is the average velocity $u'(x,t)$ is the speed fluctuating, through Navier-Stokes equations it's possible to obtain the Reynolds averaged equations. The equations for the mean motion obtained are similar to Navier-Stokes equations with the exception of the divergence of the stress tensor Reynolds: the system resulting from the Navier-Stokes equations is closed, while the system resulting from the RANS simulation is not open because Reynolds tensor introduces 6 additional unknowns. The problem mentioned is known as the problem of closure of turbulence which is solved by introducing models for the turbulent fluctuations which have to reproduce the action of fluctuating terms on mean motion.

The K-ε model is one of the most common models of turbulence, even if it is not appropriate in the case of strong adverse pressure gradients. It is a model with two equations: it includes two additional transport equations to represent properties of the turbulent flow and effects such as convection and diffusion of turbulent energy. The first variable transported is the turbulent kinetic energy, k. The second variable transported is the turbulent dissipation, ε; the second variable determines the scale of turbulence, while the first variable k determines the energy in the turbulence. There are two formulations of the K-ε models: the standard k-epsilon model and the RNG k-epsilon model.

In the standard k-epsilon model, eddy viscosity is determined by single length scale turbulence, so the turbulent diffusion is calculated only through a specified scale, whereas in reality all scales of motion will contribute to turbulent diffusion.

The approach RNG (Re-Normalisation Group), a mathematical technique that can be used to obtain a model similar to the k-epsilon turbulence, presents a modified equation ε, which attempts to explain the different scales of turbulence through changes at the term of production of turbulence. The equations used are:

a)   Kinematic Eddy Viscosity

$$v_T = C_\mu \cdot \frac{k^2}{\varepsilon}$$

(3)

b)   Turbulence Kinetic Energy

$$\frac{\partial k}{\partial t} + U_j \cdot \frac{\partial k}{\partial x_j} = \tau_{ij} \cdot \frac{\partial U_i}{\partial x_j} - \varepsilon + \frac{\partial}{\partial x_j} \cdot \left[\left(v + \frac{v_T}{\sigma_k} \right) \cdot \frac{\partial k}{\partial x_j} \right]$$

(4)

c)   Dissipation Rate

$$\frac{\partial \varepsilon}{\partial t} + U_j \cdot \frac{\partial \varepsilon}{\partial x_j} = C_{\varepsilon 1} \cdot \frac{\varepsilon}{k} \cdot \tau_{ij} \cdot \frac{\partial U_i}{\partial x_j} - C_{\varepsilon 2} \cdot \frac{\varepsilon^2}{k} + \frac{\partial}{\partial x_j} \cdot \left[\left(v + \frac{v_T}{\sigma_\varepsilon} \right) \cdot \frac{\partial \varepsilon}{\partial x_j} \right]$$

(5)

Closure coefficient for standard k-epsilon model:

$$C_{\varepsilon 1} = 1.44, \quad C_{\varepsilon 2} = 1.92, \quad C_\mu = 0.09, \quad \sigma_k = 1.0, \quad \sigma_\varepsilon = 1.3$$

PROJECT DESCRIPTION

The purpose of this project is to improve the race performance, reducing the breaking distance and increasing the bending speed. So, we decided to intervene on the drag generated by the wing during the breaking, and also on the grip provided by downforce, function of velocity. To explain the lift, and then the downforce, reference may be made to the wing of an airplane, observing its section. The latter is asymmetric, the top has a profile longer than the bottom: when the wing moves, it separates the relative flow in two parts, so the air layers scroll faster in the top. The outflow over the wing undergoes a boost and then is aerodynamic brake for formula cars accelerated towards the tail at a higher velocity than the air under the wing, which follows a shorter path. So the two currents are reunited in the tail after a same time interval, without creating imbalances. This is not just the facts, but as a first approximation, we can refer to this model. In reference to the Bernoulli trinomial law, since in the lower flow velocity is lower than in the upper, the pressure under the wing has to be greater than that above the wing. Therefore, the difference between the two pressures generates a resultant directed upwards, that is the lift, which holds the aircraft in the air. In detail, lift can be expressed as:

$$F_l = 1/2 \, \rho V^2 A C_l \cos \alpha \tag{6}$$

where:

- ρ is the medium density;
- V is the air velocity;
- A is the reference surface;
- C_1 is a lift dimensionless coefficient;
- α is the wing angle of attack.

In racing cars, the wing is mounted upside down and the vertical thrust towards the ground (downforce): this is correlated to the tires grip coefficient. The running resistance depends on its front section, its forward speed, the density of the medium and a drag coefficient. The drag coefficient (C_d) depends on the object shape and size of the object, the medium density and viscosity, the surface roughness, and the object velocity. The aerodynamic resistance (in general fluid dynamics), or drag, is related to a large number of factors, as shown by the formula:

$$F_d = 1/2 \, \rho V^2 A C_d \cos \alpha \tag{7}$$

where:

- ρ is the medium density of the;
- V is the air velocity;
- A is the reference surface (in case of aircraft is the wing surface, the car front surface);
- C_d is a drag dimensionless coefficient;
- α is the wing angle of attack.

The overall resistance opposed by a fluid medium to the object forward movement is given, in first approximation, by the sum of the frictional resistance, the wake resistance and the induced resistance of lift. In particular, for a tapered body, the flow resistance is given by friction (laminar and/or turbulent), that is the rubbing of the surface against the medium. For this purpose we introduce the concept of boundary layer: it's the dynamic range, laminar or turbulent, in which internal current speed is subject to strong gradients (continuous changes), due to the viscosity of the fluid. It can be considered as the area that undergoes a disorder, and the velocity is zero on the layer surface (Figure 3).

The thickness of the boundary layer is very small, and it is of one order of magnitude lower than the overall dimensions of the object, that generates the viscose perturbation. Then, inside the boundary layer, the tangential shear stress is "dense". For this reason in the layer is exerted an intense dissipative braking action, converting part of the movement in thermal agitation. The dissipative action limits the relative velocity between the object and the fluid, which surrounds it. In a turbulent boundary layer, the viscous stresses are added also the stresses, due to the exchange of transverse momentum; these actions increase with the fluid density. The chaos of the turbulent fluid motions implies higher thermal dissipation, so the braking opposing force, in turbulent flow conditions, is greater than that of the laminar regime. The resistance generated, in this way, is affected by the surface roughness: moreover, the rougher surfaces ignite earlier and more easily the turbulent condition in the flow, and then, determine higher resistances. Therefore, it was decided to design some ducts, on the pressure side of the wing, initially covered by special sliding plates, for increasing the aerodynamic drag and downforce [2] .

WING DESIGN

The first phase of the design is to draw the profile of the wing with a CAD software. In this way, it is possible to make a CFD simulation, to evaluate the aerodynamic performance of the wing, in terms of downforce and drag, and estimate the useful angles of attack before stall phenomenon occurs [3]

. In fluid dynamics the stall is a reduction of the lift coefficient due to an increase of the angle of attack or due to the incident velocity decrease on an aerodynamic profile, such as an airfoil, a propeller blade or a turbomachinery rotor. The minimum value of the angle of attack for which the stall occurs is called critical angle of attack. This value which corresponds to the maximum lift coefficient, changes significantly, depending on the particular profile or on the considered Reynolds number [4] . Similarly, the profile of the active cavities has been reported, and appropriate simulations were performed. In this way it was possible to estimate the sizes and configurations to achieve the project target. Based on the data collected, the application of these cavities on the wing is studied, evaluating the performance on the different possible arrangements of these cavities. At this moment only 2D simulations have been performed, and a 3D series is considered as future improvement of the project. The models, the different configurations and the results obtained from all the cases mentioned above, will be shown in detail in the following paragraphs.

GEOMETRY MODELING

To approximate the operating conditions of the wing, a control conduit with the dimensions shown in Figure 4(a) has been chosen. Regarding to the active cavity, the geometry is illustrated in Figure 4(b). The space surrounding the geometry of the aerodynamic and the cavities was discretized using a special dedicated software available as ANSYS package. Furthermore, to observe the progress of the boundary layer, it was built on a reference mesh of 5 layers, with growth factor 1.1, starting from the adjacent profiles of height 0.18 mm (Figure 5). To this purpose, a sizeable set of data was created by means of sufficiently accurate numerical simulations, to derive initial values.

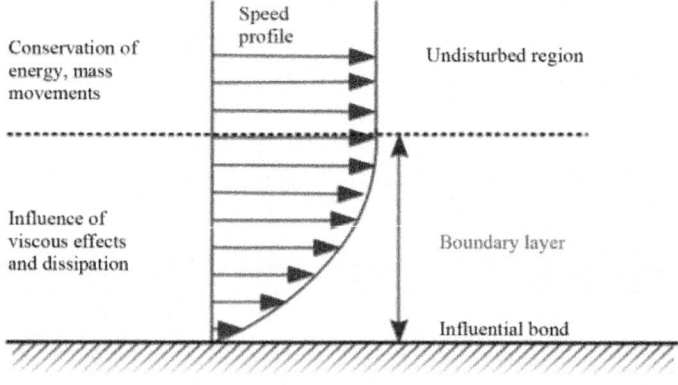

Figure 3: Boundary layer.

The simulations were performed on 3-D models in kinematic similarity using a commercial CFD simulation code, ANSYS/Fluent. The turbulence model was the k-ε realizable, with second order accuracy. Each model was meshed to ensure a $y^+_{max} \sim 5$, a necessary condition for adopting the enhanced wall treatment, since the quality of the grid has a relevant importance on the accuracy and stability of the numerical simulation.

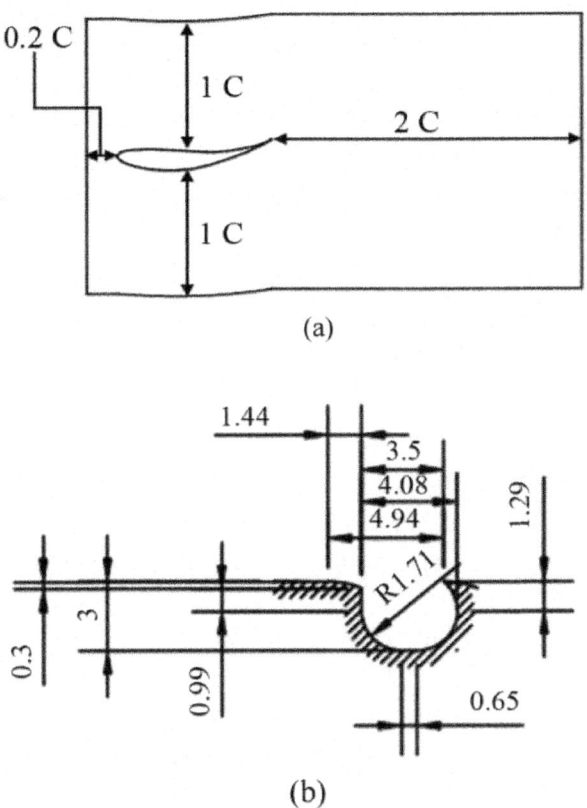

(a)

(b)

Figure 4: (a) Control conduit; (b) Active cavity dimensions (in mm).

Commercial software allows the "plastering" of cell layers to the critical boundaries of the control volume, which are obviously, in this case, the wall surfaces of the hub, casing and blades. In these zones the usual practice is that of creating a completely structured boundary layer, specifying whenever possible both the height of the first row of cells and the "growth ratio", i.e. the rate that determines the height of the successive cells. In this process, the height of the first row of cells is usually determined via an empirical formula that gives the value of a wall-based local Reynolds number, denoted by y^+

($y^+ = u^* \cdot y/v$ where $u^* = (\tau_{wall}/\rho)^{1/2}$, with τ_{wall} being the wall shear stress). For the wing analysis control volume was split in several smaller sub-volumes, to achieve a more consistent set of faces and to better exploit the possibility of creating a locally more refined grid. The choice of the boundary conditions was made as follows: it was performed heuristically, starting from the preliminary sizing data, calibrating them by means of a first simulation, adjusting the values by iteratively resetting the outlet static pressure on the near-wake radial area downstream of the trailing edge. Through subsequent simulations the values of the inlet total pressure and temperature were refined as well in order to ensure conservation of the mass flow rate (the so-called "mass flow inlet condition" was adopted). The turbulent parameters were the turbulence intensity $I = (\sqrt{k})/U$. Rotational periodicity was imposed on all lateral channel surfaces. The number of cells is about 65,000 elements. Finally, the starting boundary conditions are:

- fluid: it is considered air as an ideal gas at constant viscosity;
- input data: the pressure of 101325 Pa and temperature of 300 K represent the operating conditions.

boundary conditions:

- inlet → mass flow rate;
- outlet → pressure outlet;
- for both, the conditions relating to the model were set on intensity and length scales, with values of 5% and 0.03 m ($\simeq$ 1/10 of the rope), respectively;
- on the upper and lower walls of the duct it has set the periodicity condition;
- for wing, are set on the condition stationary wall and no slip;
- for the solution a simple high order term and relaxation has been chosen, by setting for all variables a relaxation factor of 0.25.

SIMULATION RESULTS

WING PERFORMANCE

In this paragraph, the performance of the wing has been analyzed. The results obtained by using CFD simulation (see Figures 6-8), were used as the reference model for the subsequent tests [5]. Since, the following figure shows the results for a 0° angle of attack. In particular, for a hypothetical unitary extension wing (1 m), it is obtained:

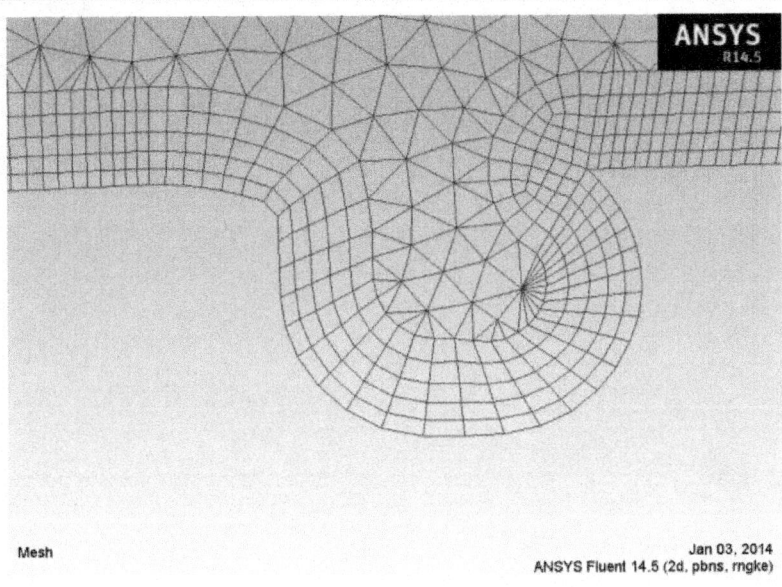

Figure 5: Cavity boundary layer mesh.

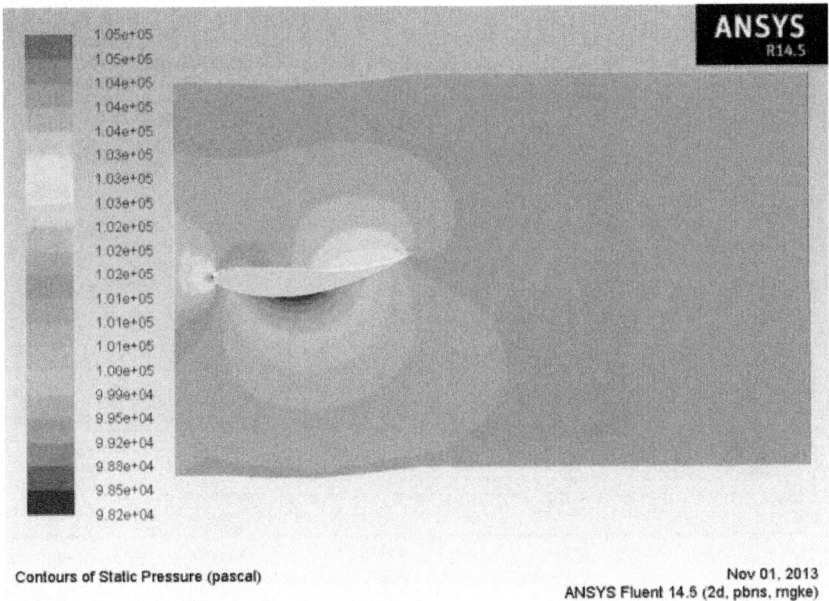

(a)

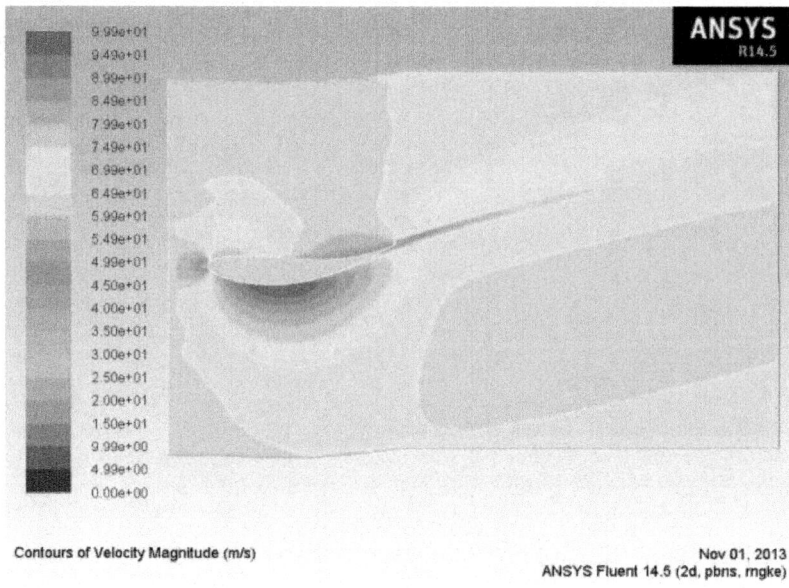

Contours of Velocity Magnitude (m/s)

Nov 01, 2013
ANSYS Fluent 14.5 (2d, pbns, rngke)

(b)

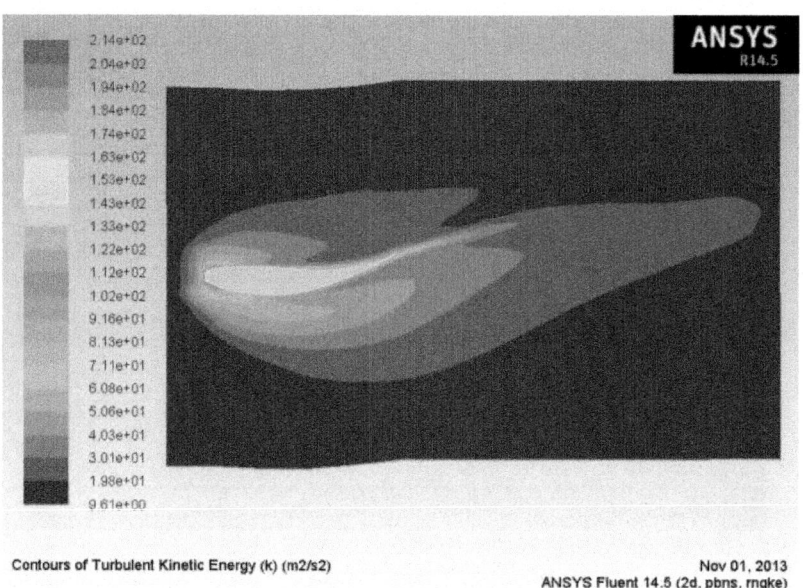

Contours of Turbulent Kinetic Energy (k) (m2/s2)

Nov 01, 2013
ANSYS Fluent 14.5 (2d, pbns, rngke)

(c)

Figure 6: Pressure, velocity and turbulence plots for 0° of angle of attack.

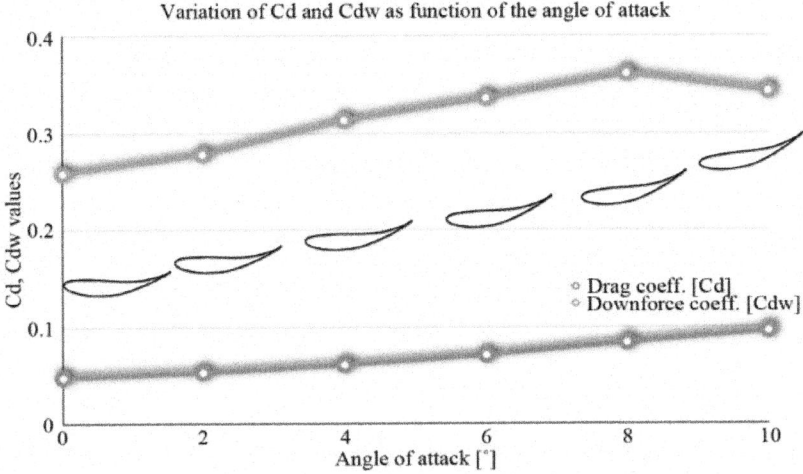

Figure 7: Plot of C_d and C_{dw} variation as function of the angle of attack.

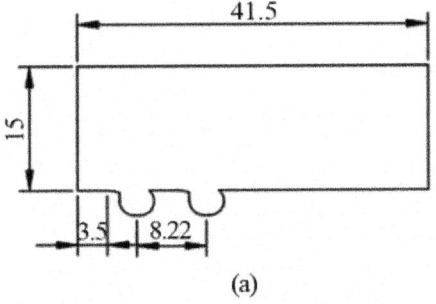

(a)

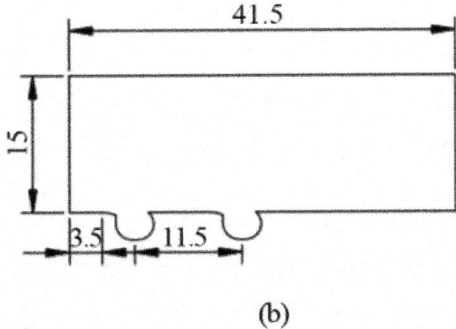

(b)

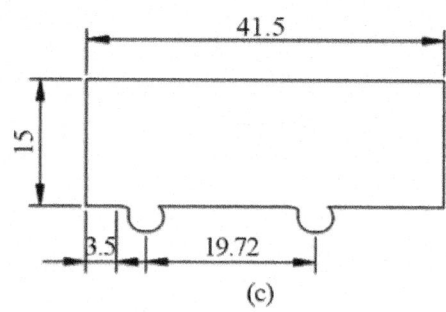

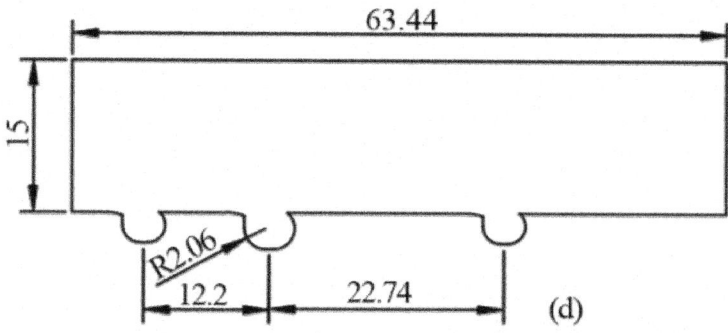

Figure 8: Different cavities configurations. (a) two cavities, (b) two cavities with space increase, (c) see case (b) (d) three cavities.

- lift coefficient $C_l = -0.25762$;
- lift $F_l = -312.66$ N;
- drag coefficient $C_d = 0.04817$;
- drag $F_d = 58.52$ N.

Finally, to individuate the stall angle, additional simulations, at different angles of attack, have been performed; precisely $2°, 4°, 6°, 8°$ and $10°$, reporting the values obtained in Table 1. It can be noticed that stall occurs for angles of attack greater than $8°$. The plot of C_d and C_{dw} is shown inFigure 9.

Active Cavity Performance Analysis

After some preliminary tests, the dimensions, that characterize the opening and depth of cavities, are 3.5 mm and 3 mm respectively. Later, built a control duct, additional simulation to find the best configuration has been carried out. In particular the configuration used were: two cavities located at 8.22 mm of distance each other, 11.5 mm, 19.72 mm and, finally, three cavities, two equal and one larger radius, interposed to 12.2 mm and 22.74 mm with respect to the remaining two (Figure 8). The analysis have enlightened that the quasi-optimal configuration is the one with the two equal cavities located at intermediate distance: in fact, the last configuration corresponds to the distance limit beyond which the reabsorption of the bubble pressure generated by the cavity in front occurs (circled in red in Figure 9). Based on these considerations, it can proceed to make the cavities on the wing to get the configurations that produce the most desired aerodynamic effects. The trend of fluid velocity is reported in Figure 10.

Wing with Front Cavities

According to the pressures pattern observed in the case of rear wing with zero angle of attack, in the first instance it is thought to have a host of ducts on the front half of the wing, because the pressure on that section is lower than the remain aerodynamic one. After that, the CAD geometry has been modified reproducing the chosen configuration, and then we proceeded with the CFD simulation (results inFigure 11). The values obtained confirm the previous considerations:

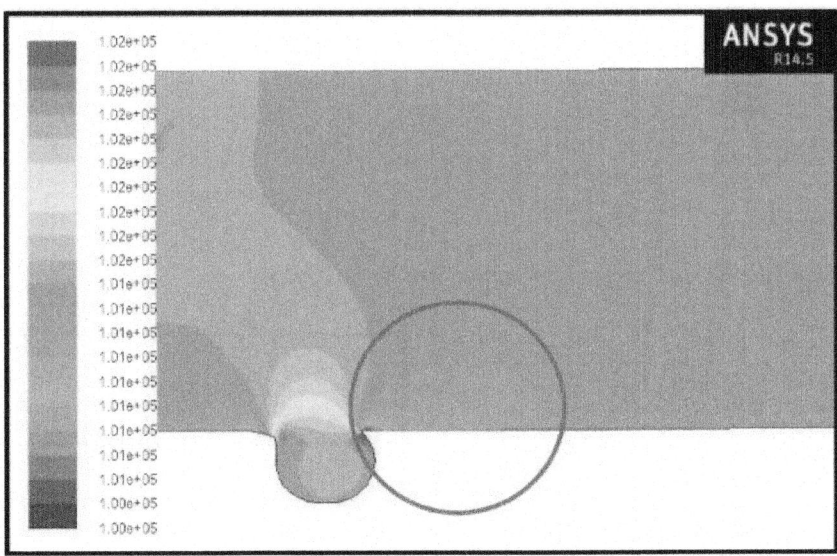

Figure 9: Pressure bubble generated by cavity. In the red circle the reabsorption of the bubble pressure generated by the cavity in front can be seen.

Table 1: Variation of C_d and C_{dw} as function of the angle of attack

Angle of attack [°]	Drag coefficient [C_d]	Downforce coefficient [C_{dw}]
0	0.048	0.258
2	0.053	0.278
4	0.061	0.313
6	0.071	0.336
8	0.084	0.361
10	0.096	0.343

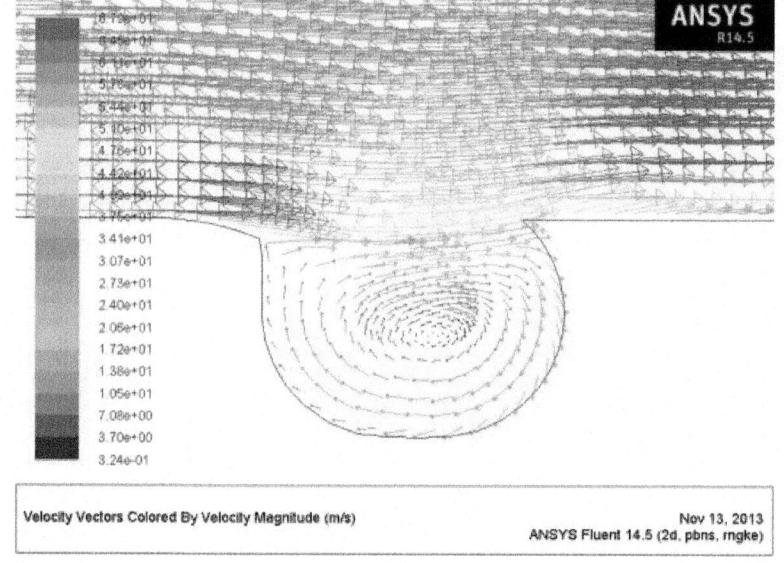

Figure 10: Velocity vectors developed inside the cavity.

$C_l = -0.26245$;
$F_l = -346.17$ N;
$C_d = 0.04777$;
$F_d = 63.00$.

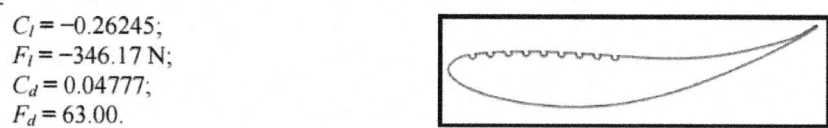

Wing with Back Cavities

For comparison, it was decided to realize the same number of cavities in the back half of the wing, leaving smooth the front half area. The CFD simulation results in Figure 12. In particular, it has that:

$C_l = -0.25834$;
$F_l = -341.12$ N;
$C_d = 0.04625$;
$F_d = 61.04$ N.

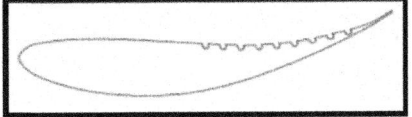

Wing with Cavities on the Whole Upper Surface

Finally (see Figure 13), it is decided to extend the group of active cavities over the entire pressure side. The coefficients and the forces derived are:

$C_l = -0.25611$;
$F_l = -365.17$ N;
$C_d = 0.04623$;
$F_d = 65.88$ N.

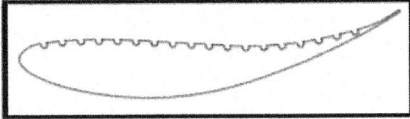

Wing with Single Cavity

For evaluating the possible influence of the single cavity on the whole group, further CFD simulations have been carried out and it is placing at different points on the back wing; appropriate pairs combinations have been also considered. Enumerating from 1 to 18 individual ducts starting from closest to the tip of the wing, the results obtained, with these additional series of simulations, are reported in Table 2.

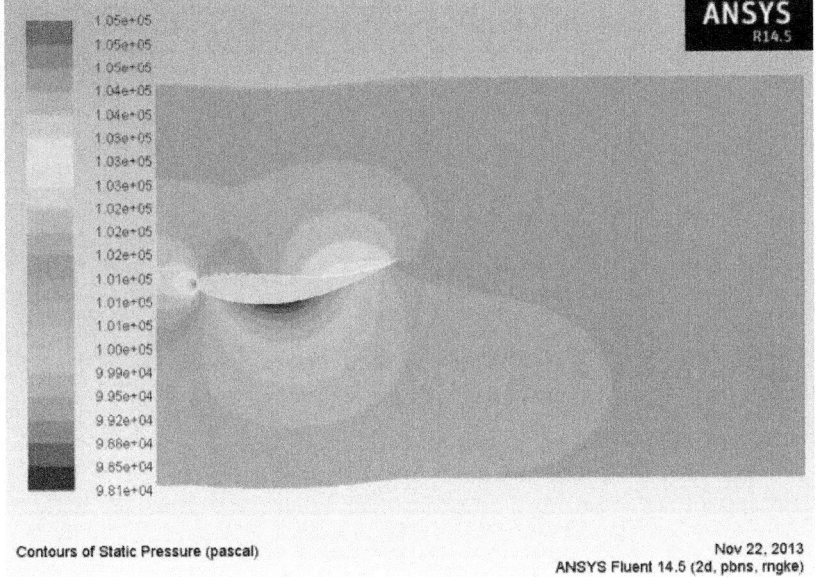

(a)

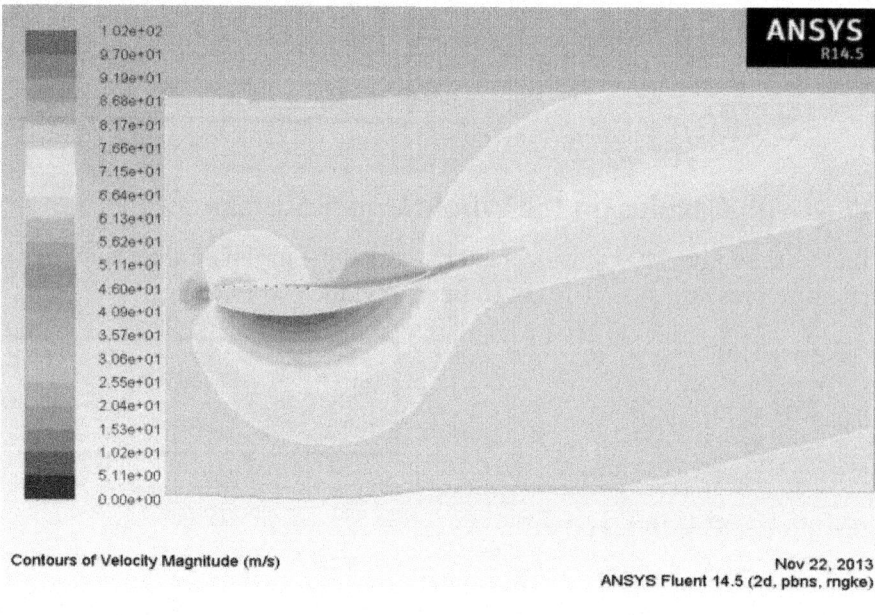

(b)

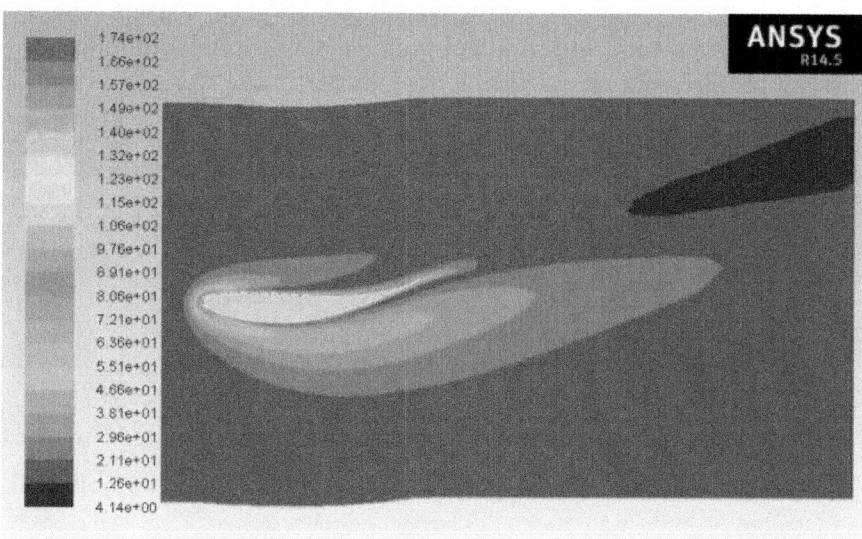

(c)

Figure 11: Pressure, velocity and turbulence plots for wing with front cavities.

It can be notice that 8 - 9 & 13 - 14 configurations denote an increase in downforce compared to the smooth wing and, at the same time, a decrease in drag.

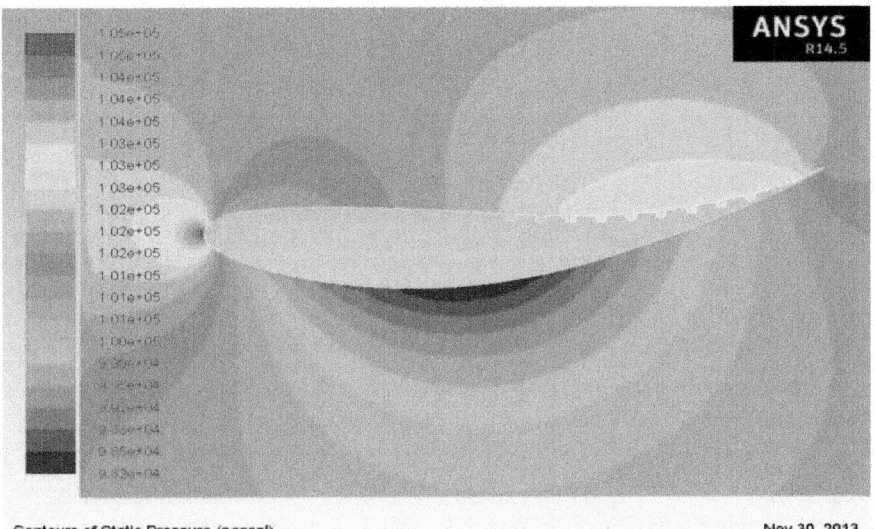

Contours of Static Pressure (pascal)   Nov 30, 2013
                                        ANSYS Fluent 14.5 (2d, pbns, rngke)

(a)

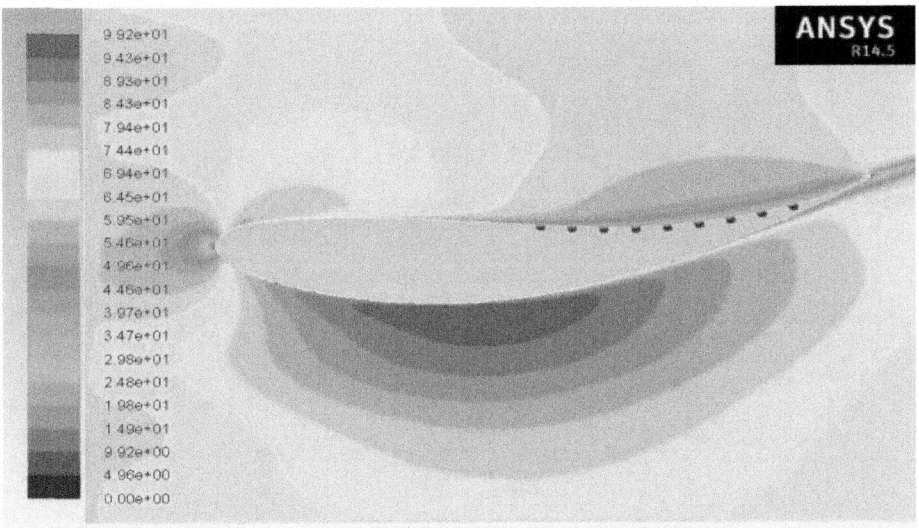

Contours of Velocity Magnitude (m/s)   Nov 30, 2013
                                        ANSYS Fluent 14.5 (2d, pbns, rngke)

(b)

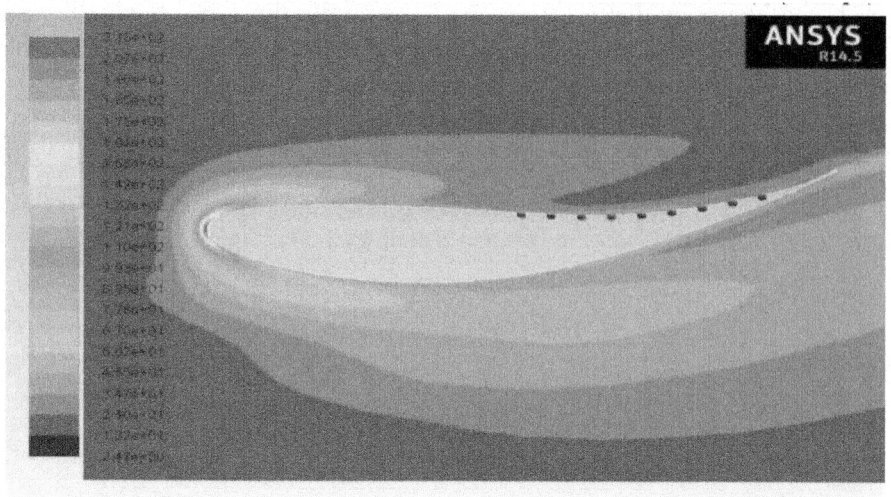

Contours of Turbulent Kinetic Energy (k) (m2/s2)   Nov 30, 2013
                                          ANSYS Fluent 14.5 (2d, pbns, rngke)

(c)

Figure 12: Pressure, velocity and turbulence plots for wing with back cavities.

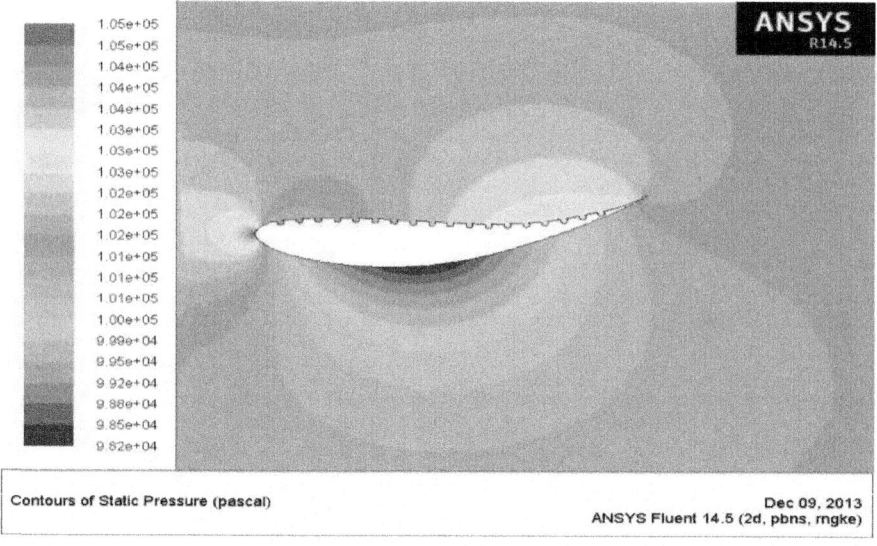

Contours of Static Pressure (pascal)   Dec 09, 2013
                                          ANSYS Fluent 14.5 (2d, pbns, rngke)

(a)

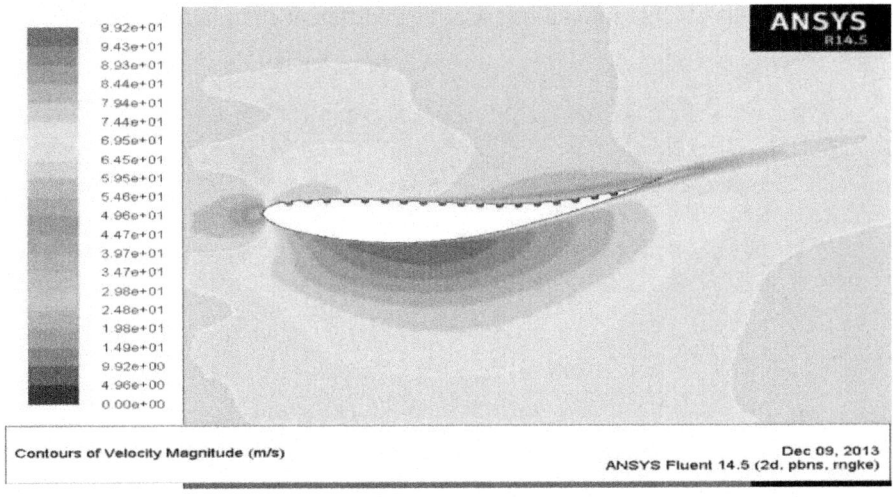

(b)

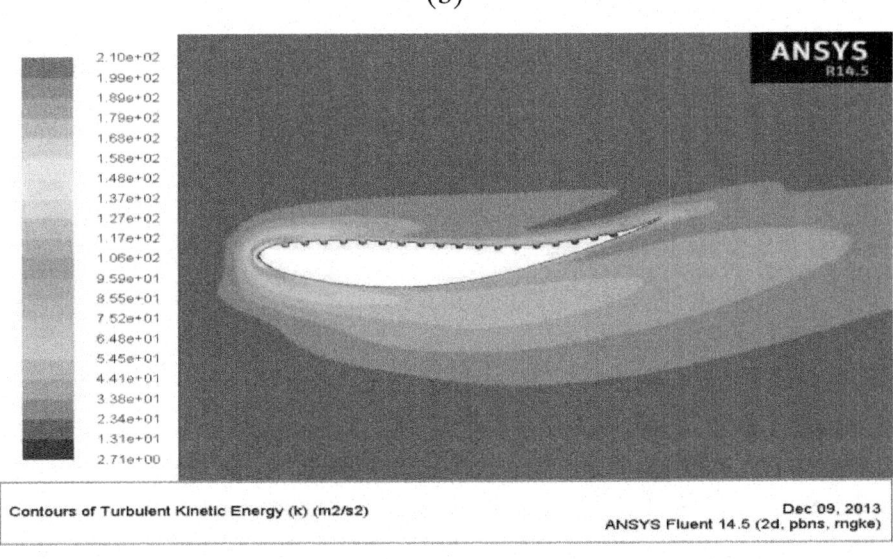

(c)

Figure 13: Pressure, velocity and turbulence plots for Wing with cavities on the whole upper surface.

Table 2: Results of CFD simulations relative to different configuration of cavities over the wing

Cavities	Reference	C_l	F_l	C_d	F_d
3		−0.25920	−317.46	0.04639	56.82
6		−0.25961	−317.96	0.04644	56.88
9		−0.25892	−317.12	0.04620	56.58
12		−0.25966	−318.01	0.04639	56.81
15		−0.25978	−318.16	0.04645	56.89
8 - 13		−0.25954	−320.88	0.04345	53.72
9 - 14		−0.25978	−321.11	0.04348	53.74
8 - 9 & 13 - 14		−0.25110	−316.38	0.03898	49.11
1 - 9 & 13 - 14		−0.25691	−344.88	0.04378	58.77

SIMULATION RESULTS ANALYSIS

The CFD simulations confirm the theoretical expectations: in particular, three of the above configurations deserve deeper analysis. In detail, it can notice that the wing with active cavities extended to the whole surface and one with the ducts only placed in the tail, show, roughly, the same value of drag (respectively 65.88 N and 61.04 N): however, in the first case, it generates greater downforce (365 N vs N 341). From this observation, it can be inferred that the rear wing with 18 ducts fulfills, in the most effective manner, a better performance, during the braking. While the same wing, but with cavities only located in the back, having a slightly lower drag, could be used during the turns: in fact, referring to the performance of the vehicle, the goal is to have a high grip to deal these turns as quickly as possible. Therefore, the choice of the most efficient set (between the configurations above) depends on the type of curve (more or less fast) and the balance between the drag and downforce forces. So, equipping the wing of pairs 8 - 9 and 13 - 14, for its characteristic to generate more downforce and a lower drag than the smooth aerodynamic profile, it could be used in starting. In fact, in this situation, to have a good sprint, it is desirable to have a low aerodynamic resistance and prevent slipping of the rear wheels: this phenomenon can be achieved by exploiting the downforce generated by the wing in this configuration. The result would be a better grip. Ultimately, you should have a wing that can satisfactory answer the two situations. The configuration must be able to provide an excellent grip and a lower resistance in starting, and get a better behavior in curves. This can be achieved by equipping the wing of sliding panels that open or close the cavities, according to the request of the vehicle.

CONCLUSIONS AND POSSIBLE IMPROVEMENTS

The CFD simulations indicate the effectiveness of active cavities, practiced on a formula car rear wing, in order to achieve an aerodynamic brake. Specifically, we can assert that the configuration with the best balance between downforce and drag is that with extended ducts over the entire top surface of the aerodynamic ($F_1 = -365.172$ N, $F_d = 65.88$ N). Finally, by exploiting the selectivity of sliding panels, as previously explained, it can be realized different wing configurations, depending on the needs required by the race and the sensitivity of the driver.

As regards any improvements, to be made on the performance testing of the brake (the object of study of this paper), more CFD simulations could perform, using a vertical bulkhead on the terminal part of the wing, as well as 3D simulations. In this way, it would be possible to observe the effects of lift drag which should be minimized. Then the next step would be to achieve physically this device, to install it on a formula car and compare experimental data with those obtained by CFD simulations.

ACKNOWLEDGEMENTS

The research has been partially sponsored by the Italian Government on the framework of Formula Cars/Formula 4 project.

REFERENCES

1. 2014 Formula 4 Technical Regulations, FIA. http://www.fia.com/sites/default/files/basicpage/file/FIIII_plaquette_full_formula4.pdf

2. Gamma, F., Sciubba, E., Zingaro, D. and Farello, G.E. (2002) Fluid Dynamic Behavior of Heat Exchangers with Active Cavities: A Numerical Study. Numerical Heat Transfer Applications, 42, 385-400.

3.   Chandra, S., Lee, A., Gorrel, S. and Greg Jensen, C. (2011) CFD Analysis of PACE Formula 1 Car. PACE, 1, 1-14.   http://dx.doi.org/10.3722/cadaps.2011.PACE.1-14

4.   Prasad. A.K. and Koseff, J.R. (1989) Reynolds Number and End-Wall Effects on a Lid-Driven Cavity Flow. Physics of Fluids A: Fluid Dynamics, 1, 208-218. http://dx.doi.org/10.1063/1.857491

5.   Chen, C.-L., Chung, Y.-C. and Lee, T.-F. (2012) Experimental and Numerical Studies on Periodic Convection Flow and Heat Transfer in a Lid-Driven Arc-Shape Cavity. International Communications in Heat and Mass Transfer, 39, 1563-1571.

Chapter 4

ADAPTIVE REAL-CODED GENETIC ALGORITHM FOR IDENTIFYING MOTOR SYSTEMS

Rong-Fong Fung[1] and Chun-Hung Lin[2]

[1]Department of Mechanical & Automation Engineering, National Kaohsiung First University of Science and Technology, Kaohsiung, Taiwan

[2]Graduate Institute of Electrical Engineering, National Kaohsiung First University of Science and Technology, Kaohsiung, Taiwan

ABSTRACT

In this paper, the main objective is to identify the parameters of motors, which includes a brushless direct current (BLDC) motor and an induction motor. The motor systems are dynamically formulated by the mechanical and electrical equations. The real-coded genetic algorithm (RGA) is adopted to identify all parameters of motors, and the standard genetic algorithm (SRGA) and various adaptive genetic algorithm (ARGAs) are compared in the rotational angular speeds and fitness values, which are the inverse of square differences of angular speeds. From numerical simulations and experimental results, it is found that the SRGA and ARGA are feasible, the ARGA can effectively solve the problems with slow convergent speed and premature phenomenon, and is more accurate in identifying system's parameters than the SRGA. From the comparisons of the ARGAs in identifying parameters of motors, the best ARGA method is obtained and could be applied to any other mechatronic systems.

INTRODUCTION

The mechanical commutator of the brushless direct current (BLDC) motor [1] is replaced by electronic switches, which supply current to the motor windings as a function of the rotor position. A BLDC motor is one of the permanent magnetic synchronous motors, and has advantages of simple structure, product easily, and low cost [2] . Due to their favorable electrical and mechanical properties, high starting torque and high efficiency, the BLDC motor are widely used in most servo applications such as actuation, robotics, machine tools, and so on.

Many literatures [3] - [6] have discussed motor operations. For a quality BLDC motor, proper maintenances and applications are important, and inherently more reliable, more efficient, and with current electronics technology, more cost effective than the standard electrical fans and controllers [3] . Therefore, the BLDC motor has advantages of high performance, low noise and long lifespan. The conventional dc motor can be represented by the mathematical model. The technicians often wanted to test the performance or measure the parameters of a motor, and the ordinary differential equations of a motor were solved by Runge-Kutta method [4] [5] . However, the precise of a BLDC motor requires its accurate parameters, which can be measured or estimated. In order to identify parameters of BLDC motors, the genetic algorithms were used to search for precise parameters [6] .

Recently, due to rapid improvements in power devices and microelectronics, the field-oriented control and feedback linearization techniques have increased induction motor drives for high-performance applications possible [7] . Induction motors are the most widely used motors in industry because they are simple to build, rugged, reliable and have good self-starting capability. Many advanced algorithms have been investigated to control induction motors, especially the vector control [8] . To design controllers, nonlinear motor models must be used for identification and optimization. Several methods have been proposed to tackle the problem of induction machine parameter estimation [9] [10] .

Genetic algorithm (GA) is a searching process based on natural selection, and now is used as a tool for searching the large, poorly understood spaces that arise in many application areas of science and engineering [11] . In this method, a large set of configurations forms a population with new generations created by selection, crossover and mutation of the current population. It is hoped that this evolution process can increase the fitness value of the population to a near optimal value. However, conventional GA often has slow convergent speed and premature phenomenon in engineering applications. In order to overcome these shortcomings, many researchers have made great efforts to improve the performance [12] [13] by proposing a variety of programs, such as base on the overall fitness value, average square deviation of population and dual species sub-population etc.

In this paper, the real-coded encoding scheme of fixed length to randomly generate the initial population is used by means of roulette wheel. This standard genetic features by use of only three basic genetic operators: selection operator, crossover operator and mutation operator, simplifies the process of genetic evolution, and easy to be understood. However, the fitness value may occur slow convergent speed and premature phenomenon in a traditional real-coded genetic algorithms (RGA) [14] , and the adaptive genetic algorithm (ARGA) is proposed to solve the problems and to find a solution near to the maximum optimization of the motor system.

This paper is organized as follows. Firstly, the BLDC and induction motors' equations are established. Secondly, the algorithms in the SRGA and ARGAs are presented and discussed. Thirdly, comparisons between the SRGA and ARGA in numerical simulations and experimental results are discussed and it is concluded that the ARGAs are better than SRGA.

IDENTIFICATION BASED ON THE RGA

Standard Real-Coded Genetic Algorithm (SRGA)

The RGA [15] [16] is an optimization searching algorithm, which simulates evolution mechanism on a computer-based platform in conjunction with natural selection and genetic mechanism. The chromosomes are expressed by vectors and each element of vectors is called a gene. The standard RGA (SRGA) has stationary crossover probability and mutation probability in the evolutional process. If the crossover probability and mutation probability are stationary values, the individual will be lack of diversity in each generation.

Adaptive Real-Coded Genetic Algorithm (ARGA)

It is important that crossover probability and mutation probability are set for genetic algorithms, the improper settings will cause falling into local optimum algorithms in search and the premature convergence. Therefore, an efficient method for a fast setting is essential. For this point, a mechanism to adjust the crossover probability and mutation probability according to the algorithmic performance is considered [17] . In this paper, the multi-method ARGA for parameters' identification of the electrical fan system will be employed.

Method 1 of ARGA

In Equations (4) and (5), the crossover probability will be reduced to preserve excellent chromosomes; on the contrary it will be added to evolutionary excellent chromosomes. And then, the mutation probability will be reduced to preserve excellent chromosomes; on the contrary it will increase the diversity of the population and avoid to falling into local optimum [18] . The overall structure of method 1 of ARGA can be described as follows.

A.   Encoding: The parameters of the BLDC motor and induction motor are composed by real-coded values.

B.   Initialization: A collection of individuals is referred to as a population. A population size of 100 is used to generate final segmentation boundaries.

C.   Fitness function: The fitness function is adopted as follows.

$$\text{fitness function} = \frac{1}{\sum_{i=1}^{n}\left[\omega(i)-\omega^{*}(i)\right]^{2}}$$

(1)

where n is the total number of sampling point, $\omega(i)$ and $\omega^{*}(i)$ are the rotation speeds by using the identified and assigned parameters in Equations (14) and (15), respectively.

D.   Selection: In this stage, the expected time of an individual being selected for recombination is proportional to its fitness value relative to the rest of the population. This operation is to achieve a mating pool with the fittest individuals selected according to a probabilistic rule that allows these individuals to be mated into new populations. The selection is carried out by using the roulette wheel method.

E.   Crossover and Mutation: The crossover is the breeding of two parents to produce a single child, who has features from both parents and thus may be better or worse than either parent according to the objective function. The primary purpose of mutation is to introduce variation and help bring back some essential genetic traits, and also to avoid the premature convergence of entire feasible space caused by some super chromosomes [19] .

To reduce the premature convergence and improve convergence rate of the SRGA, the adaptive probabilities of crossover and mutation are presented in the ARGA. The probabilities of crossover P_c and mutation P_m are respectively given as follows.

$$P_{c} = P_{c}^{0} \times \left(1+\alpha\frac{\left(f_{avg}\right)^{n_{c}}}{\left(f_{max}-f_{min}\right)^{n_{c}}+\left(f_{avg}\right)^{n_{c}}}\right)$$

(2)

$$P_{m} = P_{m}^{0} \times \left(1+\beta\frac{\left(f_{avg}\right)^{n_{c}}}{\left(f_{max}-f_{min}\right)^{n_{c}}+\left(f_{avg}\right)^{n_{c}}}\right)$$

(3)

where f_{max}, f_{min} and f_{avg} are the maximum, minimum and average individual fitness, respectively, P_c^0 and P_m^0 are crossover and mutation of probabilities, respectively and α, β, n_c are coefficient factors. In this paper, $\alpha = 0.3$, $\beta = 0.2$ and $n_c = 2$ [20] are taken.

From Equations (2) and (3), it is known that the adaptive P_c and P_m vary with fitness functions. The P_c and P_m increase when the population tends to get stuck at a local optimum and decrease when the population is scattered in the solution space.

Method 2 of ARGA

The GAs have been extensively used in different domains as a type of robust optimization method. However, the GA to demonstrate a more serious question is a premature convergence problem, less capable local optimization, the late slow convergence and can't guarantee convergence to the global optimal solution and so on. In recent years, many researches [21] [22] try to improve genetic algorithms, such as improving the encoding scheme, fitness function, genetic operator design. For this reason, the ARGA is proposed with the crossover probability P_c and mutation probability P_m as follows.

$$
P_c = \begin{cases} P_{c1} - \dfrac{(P_{c1} - P_{c2})(f' - f_{avg})}{f_{max} - f_{avg}}, & f' \geq f_{avg} \\ P_{c1}, & f' < f_{avg} \end{cases}
\tag{4}
$$

$$
P_m = \begin{cases} P_{m1} - \dfrac{(P_{m1} - P_{m2})(f - f_{avg})}{f_{max} - f_{avg}}, & f \geq f_{avg} \\ P_{m1}, & f < f_{avg} \end{cases}
\tag{5}
$$

where, f_{max} is the best individual fitness, f' is the better individual fitness in every group, f_{avg} is the average fitness, and f is every individual fitness in current generation. The P_{c1} means fixed maximal cross proba-

bility; P_{c2} means fixed minimum crossover probability; P_{m1} and P_{m2} are fixed maximal and mutation probabilities, respectively. In this paper, $P_{c1} = 0.8$, $P_{c2} = 0.6$, $P_{m1} = 0.1$, and $P_{m2} = 0.01$ [23] [24] are taken.

As a result, the adaptive P_c and P_m are able to provide the optimum P_c and P_m targets at a certain solution. The improved chromosome crossover and mutation operators ensure the convergence of the GA more than the diversity of population [25] [26] .

Method 3 of ARGA

An important problem in usage of the RGA is premature convergence, and the searching process may trap in a local optimum before the global optimum is found. This section employs an ARGA which adjusts mutation probability dynamically based on average square deviation (ASD) of population fitness value, which shows the population diversity to solve the premature problem. From compared analysis, it is shown the proposed ARGA efficiently avoid the premature problem [27] . Premature convergence can also be blamed in [28] by avoiding the loss of critical alleles due to selection and the schemata disruption due to crossover.

The selection operation reduces the diversity of population, the crossover operation does not decrease the diversity of population, and the mutation operation can advance the diversity. Mainly, all these issues produce two effects, the lack of diversity in the population and a disproportionate exploitation or exploration relationship, cause the premature problem [29] . When the ASD becomes smaller or less, it shows that many individuals are becoming as the same, so the mutation probability should be increased to advance the diversity of population for getting global optimal solutions.

A.   Selection operator: The selection is carried out using the roulette wheel method in this paper.

B.   Crossover operator: The crossover operation does not decrease the diversity of population, and the crossover probability is fixed.

C.   Mutation operator: This section employs an adaptive method to adjust mutation probability dynamically based on the ASD value. When the ASD decreasing, mutation probability will be increased to advance the population diversity. The relationship between mutation probability and ASD is given as follows.

$$P_m = M_a \times \left(1 + \frac{f_{\max} - \mathrm{ASD}_t}{f_{\max} + \mathrm{ASD}_t}\right)$$

(6)

where $f_{\max}$ is the maximum individual fitness, M_a is mutation probability originally, and ASD_t is the ASD of population in the tth generation, and is described as:

$$ASD_t = \frac{1}{N}\sqrt{\sum_{i=1}^{N}\left(f_t^i - f_v\right)}$$

(7)

where f_t^i is the fitness value of the ith individual of the tth generation, f_v is the average fitness of the tth generation, and N is the number of population.

Method 4 of ARGA

In this section, a new GA with two species is proposed for the ARGA. The dual-specie GA composes of two sub-populations that constitute of same size individuals. The sub-populations have different characteristics, such as crossover probability and mutation operator. In one sub-population, the parents with higher similarity are cross with higher probability, and mutate with general mutation operator. In the other sub-population, the parents with smaller similarity are cross with higher probability and mutate with big mutation probability. Therefore, the new algorithm can obtain good exploitation and exploration ability [30] .

Multi-population GA [31] - [33] is an extension of traditional single population GA by dividing a population into several isolated sub-populations, within which the evolution proceeds and individuals are allowed to migrate from one sub-population to another [34] , and the flow chart is shown as follows (Figure 1).

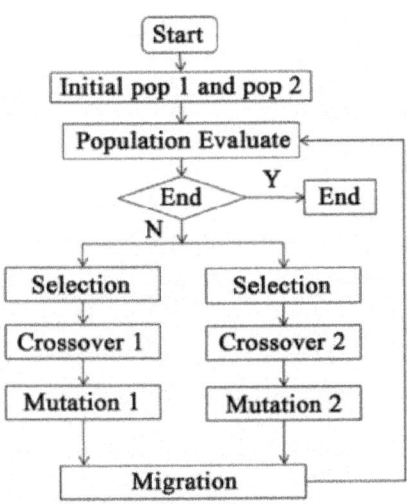

Figure 1: The GA of the dual-populations.

For the dual-population GA, one has the following operators:

A. Selection operator: The proposed algorithm establishes two separate sub-populations by random initialization, and then carries on evolution inside single sub-population and migration between two sub-populations. Here, the roulette selection is employed.

B. Crossover operator: The crossover probability is correlated with parents' similarity. The similarity between two individuals is defined as:

$$S(X,Y) = 1 - \sum_i w_i dist(x_i, y_i) \tag{8}$$

$$dist(x_i, y_i) = \frac{|f_{xi} - f_{yi}|}{|f_{max} - f_{min}|} \tag{9}$$

where x_i and y_i are two individuals, $dist(x_i, y_i)$ is the distance between x_i and y_i, f_{max} and f_{min} are the maximum and minimum individual fitness values, respectively, f_{x_i} and f_{y_i} are the fitness values of x_i and y_i, respectively.

The crossover1 emphasized local search ability, and crossover probability P_{c1} is positive correlated with the parents' similarity; the crossover2 emphasized global search ability, and crossover probability P_{c2} is negative correlated with the parents' similarity. It defines:

$$P_{c1} = \begin{cases} 1, & S(X,Y) > 0.8 \\ 0.2, & S(X,Y) < 0.2 \\ S(X,Y), & \text{else} \end{cases} \tag{10}$$

$$P_{c2} = \begin{cases} 1, & S(X,Y) < 0.2 \\ 0.2, & S(X,Y) > 0.8 \\ 1 - S(X,Y), & \text{else} \end{cases} \tag{11}$$

C. Mutation operator: Mutation1 is the normal mutation which proceeds with constant probability. Mutation 2 needs to have ability to robustly explore the solution space and to escape from local peak. The probability of mutation 2 is adaptive mutation probability, and the probability value P_{m2} is considerably large and correlate with P_{m1} and the diversity of two sub-populations. The P_{m2} is defined as:

$$P_{m1} = \text{fixedmutation probability} \tag{12}$$

$$P_{m2} = \begin{cases} 2P_{m1}, & \text{if } f_{max2} - f_{avg2} > f_{max1} - f_{avg1} \\ 4P_{m1}, & \text{else} \end{cases} \tag{13}$$

where f_{max1} and f_{max2} are the maximal fitness values; f_{avg1} and f_{avg2} are the average fitness values of population 1 and population 2, respectively.

EXAMPLES

The different four ARGAs will be applied to the BLDC and induction motors. At first, it is needed to show the governing equations of the motor system, and find out what are parameters to be identified.

Equation of a BLDC Motor

The BLDC motor [35] is one kind of permanent magnet synchronous motor, and has permanent magnets on the rotor and trapezoidal-shape back EMF. The BLDC motor employs a dc power supply switched to the stator phase windings of the motor by power devices, and the switching sequence is determined from the rotor position. The phase current of BLDC motor, in typically rectangular shape, is synchronized with the back EMF to produce constant torque at a constant speed.

A commonly used second-order linear model for a BLDC motor [36] can be expressed mathematically as

$$L_a di_a / dt + R_a i_a + K_b \omega = v_a \tag{14}$$

$$J_m d\omega / dt + B_m \omega + T_f = K_t i_a \tag{15}$$

where L_a is armature inductance, i_a is motor armature current, R_a is resistance, K_b is back-EMF constant, ω is angular speed of the motor shaft, v_a is armature voltage: J_m is inertia of the motor, B_m is viscous damping coefficient, T_f is the frictional torque, and K_t is torque constant of the motor. In Equations (14) and (15), a the rotational speed and electric current are the state variables and the electrical voltage is an input.

Equation of an Induction Motor

The field-oriented induction motor drive can be applied for high-performance industrial applications, and the controllers implemented in induction motor drives are generally based on the system mathematical model. The parameter identification in a rotation rotor is very useful in monitoring and testing a high-power induction motor drive, and then its performance depends heavily on the motor parameters [37] [38] . In the decoupling condition, main parametric uncertainties of induction motors are the mechanical parameters and load torque disturbances, which are slowly time-varying in general [39] . Measurements of the rotational angular speeds and input electrical voltages are required for the system identification procedure [40] . Model-based methods of rotation-speed estimation are characterized by their simplicity, but sensitivity to parameter variations is considered as the major problem [41] .

The complete electrical and mechanical models [42] are combined, and its electro-mechanical equation can be expressed as follows:

$$
\frac{d}{dt}\begin{bmatrix} i_{ds} \\ i_{qs} \\ \lambda_{dr} \\ \lambda_{qr} \\ \omega_r \end{bmatrix} =
\begin{bmatrix}
\dfrac{-R_s}{\sigma L_s} - \dfrac{R_r(1-\sigma)}{\sigma L_r} & \omega_e & \dfrac{L_m R_r}{\sigma L_s L_r^2} & 0 & 0 \\
-\omega_e & \dfrac{-R_s}{\sigma L_s} - \dfrac{R_r(1-\sigma)}{\sigma L_r} & 0 & \dfrac{L_m R_r}{\sigma L_s L_r^2} & 0 \\
\dfrac{L_m R_r}{L_r} & 0 & -\dfrac{R_r}{L_r} & \omega_e & 0 \\
0 & \dfrac{L_m R_r}{L_r} & -\omega_e & -\dfrac{R_r}{L_r} & 0 \\
0 & 0 & 0 & 0 & -\dfrac{B_m}{J_m}
\end{bmatrix}
\begin{bmatrix} i_{ds} \\ i_{qs} \\ \lambda_{dr} \\ \lambda_{qr} \\ \omega_r \end{bmatrix}
$$

$$
+ \begin{bmatrix}
\dfrac{n_p L_m}{\sigma L_s L_r}\omega_r \cdot \lambda_{qr} \\
-\dfrac{n_p L_m}{\sigma L_s L_r}\omega_r \cdot \lambda_{dr} \\
-n_p \omega_r \cdot \lambda_{qr} \\
n_p \omega_r \cdot \lambda_{dr} \\
\dfrac{3 L_m}{2 L_r} n_p \left(i_{qs}\lambda_{dr} - i_{ds}\lambda_{qr} \right)
\end{bmatrix}
+ \begin{bmatrix}
\dfrac{v_{ds}}{\sigma L_s} \\
\dfrac{v_{qs}}{\sigma L_s} \\
0 \\
0 \\
-\dfrac{T_L}{J_m}
\end{bmatrix},
\tag{16}
$$

where i_{ds} is the d-axis stator current, i_{qs} is the q-axis stator current, λ_{dr} is the d-axis rotor flux linkage, λ_{qr} is the q-axis rotor flux linkage, ω_r is the angular speed of the rotor, R_s is the stator resistance, L_s is the stator inductance, R_r is the rotor resistance, L_r is the rotor inductance, ω_e is the electrical angular speed, n_p is the number of pole pairs, L_m is the magnetizing inductance, B_m is the motor damping coefficient, J_m is the motor moment inertia, v_{ds} is the d-axis stator voltage, v_{qs} flux linkage is the q-axis stator voltage and T_L is the load torque. In equation (16), the rotational angular speed, electric currents and flux linkages are the state variables and the electrical voltages (v_{ds}, v_{qs}) are the inputs.

NUMERICAL SIMULATION

For the BLDC Motor

In the numerical simulations, the input voltage is defined as follows:

$$v_a(t) = \frac{v_0}{T_1}t, \ 0 \leq t < T_1 \tag{17}$$

$$v_a(t) = v_0 + v_1 \sin \omega t, \ T_1 \leq t < T_p \tag{18}$$

where $\omega = \omega_0 + (\omega_1 - \omega_0)t/T_p$, ω_0 and ω_1 represent the minimum and maximum frequencies, respectively. V_0 and V_1 are the base voltage and bias amplitude. T_1 is the time for increasing voltage and T_p is the total time. In this paper, $v_0 = 12$ V, $v_1 = 2$ V, $\omega_0 = 5$ rad/sec, $\omega_1 = 20$ rad/sec, $T_1 = 1.5$ sec and $T_p = 10$ sec are taken. The input voltage is shown in Figure 2(a).

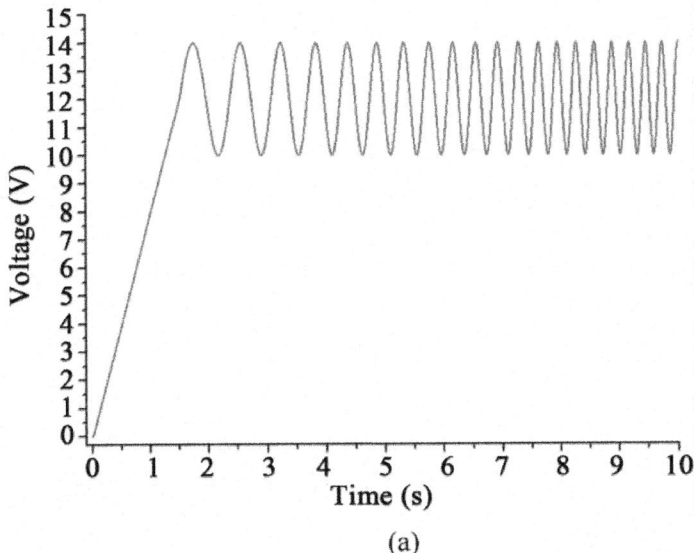

(a)

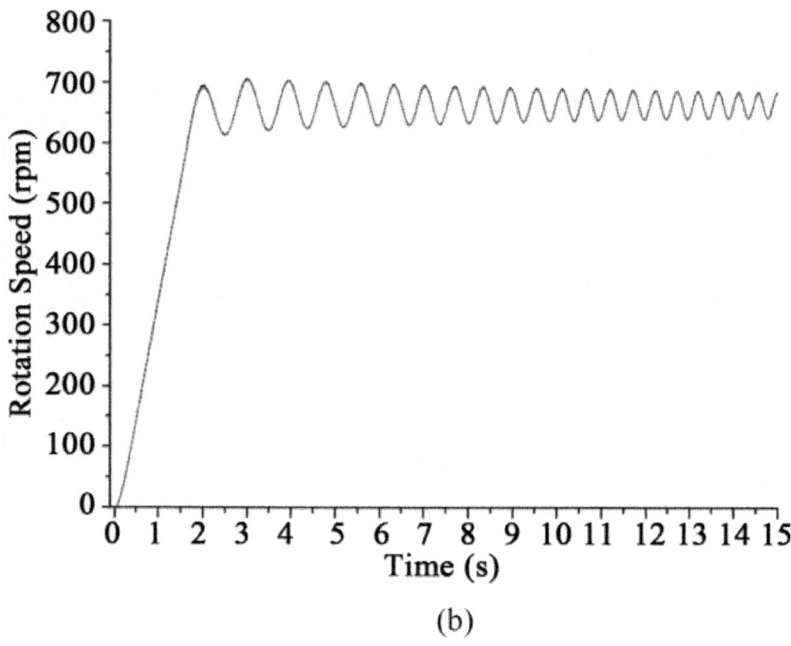

(b)

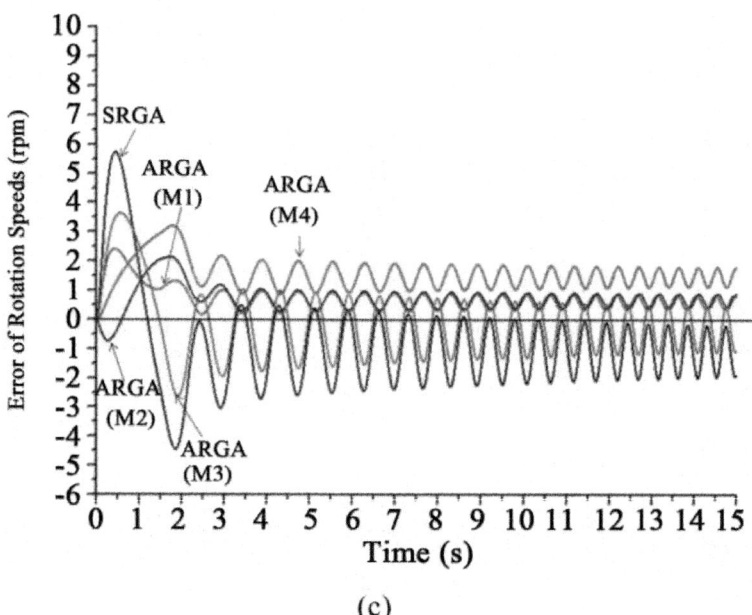

(c)

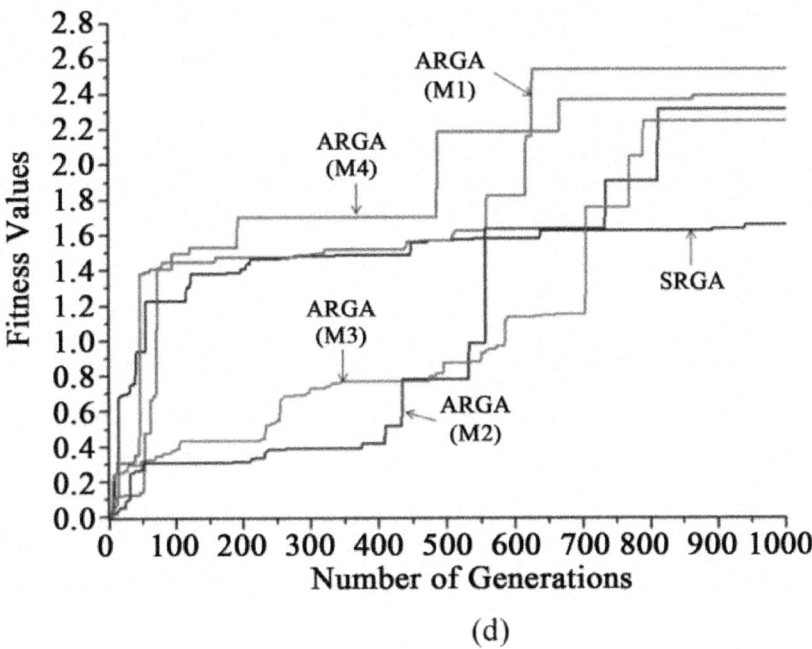

(d)

Figure 2: The comparisons of the ARGAs and SRGA. (a) The input voltage; (b) The rotation speed; (c) The errors of rotation speeds between the assigned and identified parameters; (d) The fitness values.

In order to investigate the ARGAs, compare them and find the best one for system identification of the electrical fan, the parameters $B_m = 0.001\,(\mathrm{N \cdot m \cdot s/rad})$, $J_m = 0.001\,(\mathrm{N \cdot m \cdot s^2/rad})$, $R_a = 2\,(\Omega)$, $L_a = 4\,(\mathrm{H})$, $K_b = 0.15\,(\mathrm{V/rpm})$, $K_t = 0.15\,(\mathrm{N \cdot m/A})$ and $\alpha = 10^{-5}\,(\mathrm{m \cdot N \cdot m/rpm^2})$ are assigned. Substituting these parameters into Equations (14) and (15) and using the electrical input voltages (17, 18), the rotation speed is obtained as shown inFigure 2(b).

From the electrical input voltages and rotation output speed, by using the ARGAs and SRGA the identified parameters and fitness values can be obtained as shown in Table 1. It is seen that the fitness value is largest for method 1 of ARGA, and is smallest for the SRGA. The errors of rotation speeds, which are solved by using the assigned parameters and identified parameters as shown inTable 1, are shown in Figure 2(c). It is seen that the errors of rotation speeds are smaller by the ARGAs than that by the SRGA. The fitness values of the ARGAs and SRGA are compared in Figure 2(d), it is seen that the fitness value of multi-ARGA are not only larger than SRGA, but also the errors of rotation speed are smaller in Figure 2(c).

From comparisons in Figure 2(d), it is demonstrated that the ARGAs are more efficient to identify system's parameters than the SRGA in generation numbers. It means that the ARGAs do not fall into local optimum and prevent the premature convergence, and the fitness values of ARGAs are bigger than the SRGA. The number simulations are compared,and the fitness value of method 1 of ARGA is 2.54. It is the biggest value than the other ARGAs. Moreover, its error percentages are small not only in parameters but also in the rotation speeds. That means that the method 1 of ARGA is the best algorithm to identify system's parameters.

For the Induction Motor

In an induction motor, any vector in a rotating coordinate can be described as follows:

$$f_s = f_{as}e^{j(-\theta)} + f_{bs}e^{j\left(\frac{2}{3}\pi-\theta\right)} + f_{cs}e^{j\left(\frac{4}{3}\pi-\theta\right)} = f_{ds} + jf_{qs} \tag{19}$$

According to Euler's formula, the three-phase part can be rewritten as:

$$f_s = f_{as}\left[\cos(-\theta)+j\sin(-\theta)\right] + f_{bs}\left[\cos\left(\frac{2}{3}-\theta\right)+j\sin\left(\frac{2}{3}-\theta\right)\right]$$
$$+ f_{cs}\left[\cos\left(\frac{4}{3}-\theta\right)+j\sin\left(\frac{4}{3}-\theta\right)\right] \tag{20}$$

Therefore, the relation formula can be obtained as follows:

Table 1: The assigned and identified parameters of a BLDC motor by numerical simulations

Parameters	Assigned	Feasible Domains	Identified Values / Error Percentages				
			SRGA	ARGA (M1)	ARGA (M2)	ARGA (M3)	ARGA (M4)
$R_s\ (\Omega)$	2	0 - 4	2.78/39.90%	1.92/4.20%	1.49/25.37%	1.53/23.41%	1.77/11.59%
$L_s\ (\text{H})$	4	0 - 8	3.64/8.91%	4.11/2.85%	3.94/1.41%	3.59/10.32%	4.80/19.88%
$J_m \times 10^3\ (\text{N}\cdot\text{m}\cdot\text{s}^2/\text{rad})$	1	0 - 2	0.80/20.50%	0.68/31.60%	0.52/47.90%	0.69/31.50%	0.39/60.90%
$B_m \times 10^3\ (\text{N}\cdot\text{m}\cdot\text{s}/\text{rad})$	1	0 - 2	1.18/18.00%	1.00/0.00%	1.02/1.90%	1.23/23.40%	0.81/19.90%
$K_t\ (\text{N}\cdot\text{m}/\text{A})$	0.15	0 - 0.3	0.14/8.80%	0.15/1.05%	0.16/3.79%	0.15/2.15%	0.16/4.15%
$\alpha \times 10^5\ (\text{m}\cdot\text{N}\cdot\text{m}/\text{rpm}^2)$	1	0 - 2	0.80/20.00%	1.00/0.00%	1.10/10.00%	1.00/0.00%	0.90/10.00%
Fitness value			1.658	2.540	2.314	2.249	2.391
Error % of Rotation Speed			0.51%	0.20%	0.29%	0.31%	0.30%
Convergence Generation			994	626	812	790	864

$$f_{ds} = f_{as}\cos(-\theta) + f_{bs}\cos\left(\frac{2}{3}\pi-\theta\right) + f_{cs}\cos\left(\frac{4}{3}\pi-\theta\right) \tag{21}$$

$$f_{qs} = f_{as} \sin(-\theta) + f_{bs} \sin\left(\frac{2}{3}\pi - \theta\right) + f_{cs} \sin\left(\frac{4}{3}\pi - \theta\right)$$

(22)

The stator transformation formula between the three-phase coordinate and d-q axis is shown as follows:

$$\begin{bmatrix} f_{ds} \\ f_{qs} \\ f_{0s} \end{bmatrix} = \frac{2}{3} \begin{bmatrix} \cos\theta & \cos\left(\frac{2\pi}{3}-\theta\right) & \cos\left(\frac{4\pi}{3}-\theta\right) \\ -\sin\theta & \sin\left(\frac{2\pi}{3}-\theta\right) & \sin\left(\frac{4\pi}{3}-\theta\right) \\ \frac{1}{2} & \frac{1}{2} & \frac{1}{2} \end{bmatrix} \begin{bmatrix} f_{as} \\ f_{bs} \\ f_{cs} \end{bmatrix}$$

(23)

If there is a voltage amplitude $v_m(t)$ with a frequency f, the three-phase voltage is

$$\begin{cases} v_{as} = v_m(t)\cos(2\pi f t) \\ v_{bs} = v_m(t)\cos(2\pi f t + 2/3\pi) \\ v_{cs} = v_m(t)\cos(2\pi f t + 4/3\pi) \end{cases}$$

(24)

The input electrical voltages (v_{ds}, v_{qs}) in the d-q axis are as follows:

$$\begin{bmatrix} v_{ds} \\ v_{qs} \\ v_{0s} \end{bmatrix} = \frac{2}{3} \begin{bmatrix} \cos\theta & \cos\left(\frac{2\pi}{3}-\theta\right) & \cos\left(\frac{4\pi}{3}-\theta\right) \\ -\sin\theta & \sin\left(\frac{2\pi}{3}-\theta\right) & \sin\left(\frac{4\pi}{3}-\theta\right) \\ \frac{1}{2} & \frac{1}{2} & \frac{1}{2} \end{bmatrix} \begin{bmatrix} v_{as} \\ v_{bs} \\ v_{cs} \end{bmatrix}$$

(25)

The input voltage is with alternative current (AC) and is shown as Equation (24). If the fixed frequency is defined as $f = 50$ Hz, its amplitude $v_m(t)$ is:

$$v_m(t) = \frac{v_0}{T_1} t, \quad 0 \le t < t_1$$

(26)

$$v_m(t) = v_0 + v_1 \sin\omega(t), \quad t_1 \le t < t_p$$

(27)

where $\omega = \omega_0 + (\omega_1 - \omega_0)t/t_p$, ω_0 and ω_1 represent the minimum and maximum frequencies, respectively. v_0 and v_1 are the base voltage and bias amplitude, respectively. t_1 is the total time for increasing voltage and

t_p is the total time for input voltages. In this paper, $v_0 = 90$ (V), $v_1 = 20$ (V), $\omega_0 = 5$, $\omega_1 = 20$, $t_1 = 2$ sec and $t_p = 15$ sec are assigned. The input voltage is shown in Figure 3(a).

In order to investigate the ARGAs for system identification of the electrical fan, the parameters $R_s = 0.83\ (\Omega)$, $R_r = 0.53\ (\Omega)$, $L_s = 0.086\ (\text{H})$, $L_r = 0.086\ (\text{H})$, $L_m = 0.082\ (\text{H})$, $J_m = 0.033\ (\text{N}\cdot\text{m}\cdot\text{s}^2/\text{rad})$ and $B_m = 0.055\ (\text{N}\cdot\text{m}\cdot\text{s}/\text{rad})$ are assigned. Substituting these parameters into Equation (16) and using the electrical input voltages (26, 27), the rotation speeds are obtained and shown in Figure 3(b).

From the electrical input voltages and rotational output speed, the identified parameters and fitness values by using the ARGAs and SRGA can also be obtained and shown in Table 2. It is seen that the fitness value is the biggest one for method 1 of ARGA, and is the smallest one for the SRGA. The errors of rotation speeds, which are solved by using the assigned parameters and identified parameters as shown in Table 2, are compared in Figure 3(c). It is seen that the errors of rotation speeds are smaller by the ARGAs than that by the SRGA. The fitness values of the ARGAs and SRGA are compared in Figure 3(d), it is seen that the fitness values of ARGAs are not only larger than the SRGA, but also the errors of rotation speed are smaller in Figure 3(c).

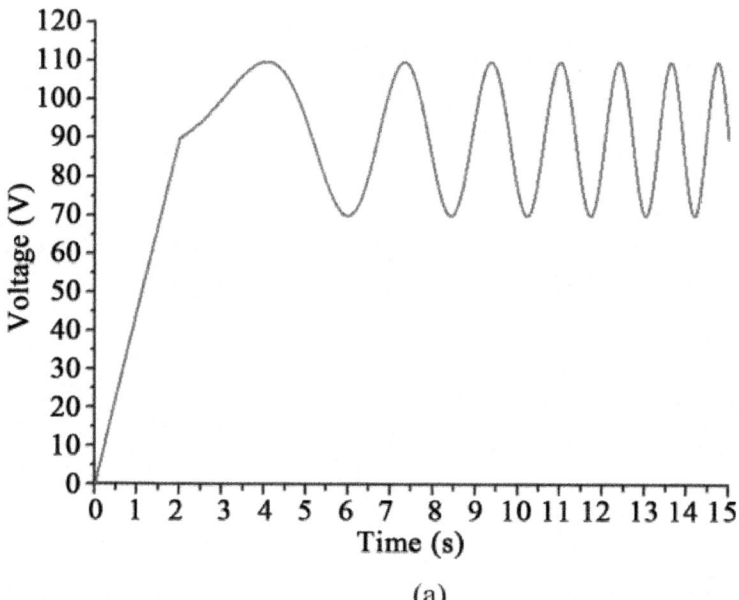

(a)

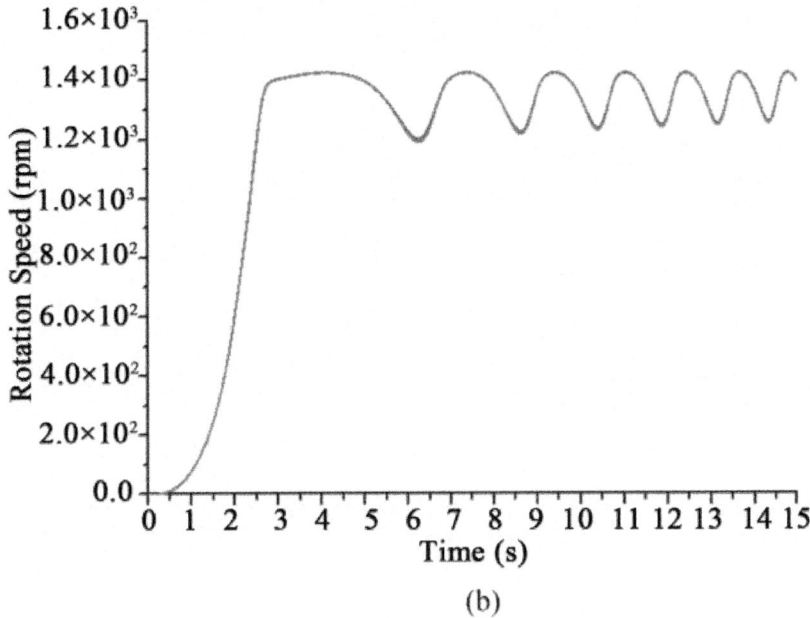

(b)

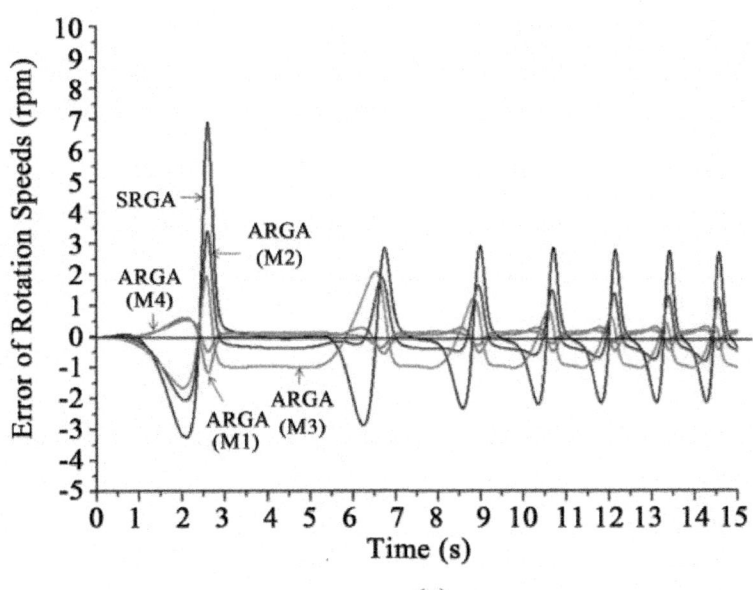

(c)

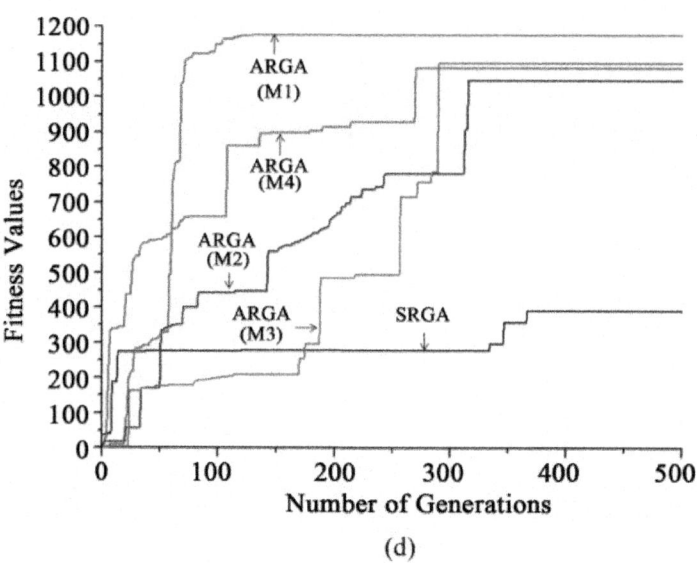

(d)

Figure 3: The comparisons of the ARGAs and SRGA. (a) The input voltage; (b) The rotation speed; (c) The errors of rotation speeds between the assigned and identified parameters; (d) The fitness values.

Table 2: The assigned and identified parameters of an induction motor by numerical simulations

	Assigned	Feasible Domains	Identified Values/Error Percentages				
			SRGA	ARGA (M1)	ARGA (M2)	ARGA (M3)	ARGA (M4)
$R_s \times 10^2$ (Ω)	83	82 - 84	82.41/0.71%	83.30/0.36%	83.61/0.74%	82.74/0.31%	83.64/0.77%
$R_r \times 10^2$ (Ω)	53	52 - 54	52.30/1.32%	52.00/1.88%	53.87/1.54%	52.17/1.56%	53.99/1.86%
$L_s \times 10^3$ (Ω)	86	70 - 90	72.82/15.33%	86.51/0.59%	84.81/1.85%	86.65/0.76%	85.49/0.59%
$L_r \times 10^3$ (Ω)	86	70 - 90	75.00/12.82%	85.28/0.83%	84.85/1.34%	87.28/1.49%	84.81/1.38%
$L_m \times 10^3$ (H)	82	70 - 90	79.71/2.79%	83.11/1.35%	81.76/0.29%	79.62/2.90%	80.81/1.45%
$J_m \times 10^3$ (N·m·s²/rad)	33	20 - 40	33.96/2.91%	33.83/2.52%	34.18/3.57%	32.14/2.58%	33.25/0.77%
$B_m \times 10^3$ (N·m·s/rad)	55	40 - 60	56.40/2.55%	55.16/0.28%	54.47/0.96%	55.50/0.91%	53.95/1.91%
Fitness value	.	.	392.37	1178.70	1049.55	1098.13	1085.16
Error % of rotation speed	.	.	0.1463%	0.0248%	0.0758%	0.0956%	0.0204%
Convergence Generation	.	.	368	130	317	291	271

From comparisons in Figure 3(d), it is demonstrated that the ARGAs are more efficient to identify system's parameters than the SRGA in generation numbers. It means that the ARGAs do not fall into local optimum and prevent the premature convergence, and the fitness values of ARGAs are bigger than the SRGA. The fitness value of method 1 of ARGA is the biggest one in all the ARGAs. Moreover, its error percentages are smaller not only in parameters but also in the rotation speeds. It means that the method 1 of ARGA is the best algorithm among the four ARGAs.

EXPERIMENTAL SETUP

For the BLDC Motor

In experiments, the electrical input voltage and rotation output speed of a real motor are to be measured, and the ARGAs are to be employed to obtain system's parameters. Figure 4 shows the experimental setup, where the computer command is transformed by the driver to a BLDC fan. The input DC voltage and the rotation speed is measured and transformed by the D/A card to computer for the identification computation. The desktop computer edits C language to control microchips, and the inverter will give signal to power supplier, which drives electric fan. At first, the voltage is controlled into the microchips, and the voltage frequency is converted as a sinusoidal function. Secondly, the system is stimulated by the voltage frequency, and the signals of the rotation speeds can be obtained by inverter, and the data is the target to be identify by the ARGAs. At last, the identifiable parameters are substituted into Equations (14) and (15) to obtain the rotation speeds by C language.

Comparisons between the Experimental and Identified Results

In experimental results, three different input voltages are given to the BLDC motor for system's identification. From numerical simulations, the method1 of ARGA, which not only accurately search for parameters, but also has small generation number inquick convergence, is the best method to identify parameters, and is also applied to the experimental system.

In order to compare experimental results, three different input voltages are employed by hand. The input voltages are ascendant before 1.5 sec and the stable after 1.5 sec, and shown in Figure 5(a) with steady-state voltages: 10.6 V, 11.5 V and 12 V. The rotation speeds for various voltage are comparing in Figure 5(b), where the curves of the experimental and identified speeds are compared. In Figure 5(b), the identified rotation speeds are approximate with the experimental rotation speeds. The error percentages of rotation speeds are very small when the voltages are stable. The identified and experimental parameters are very approximate. The speed errors of the identified speeds with respect to the experimental ones are shown in Figure 5(c). For the BLDC motor system, the fitness values with respect to total number of generations is shown in Figure 5(d) and all the fitness values converge near the 650th generation.

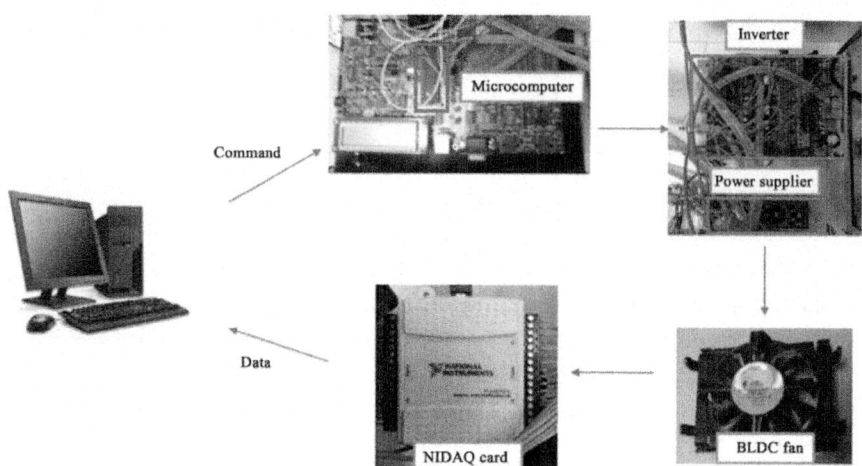

Figure 4: Experimental setup.

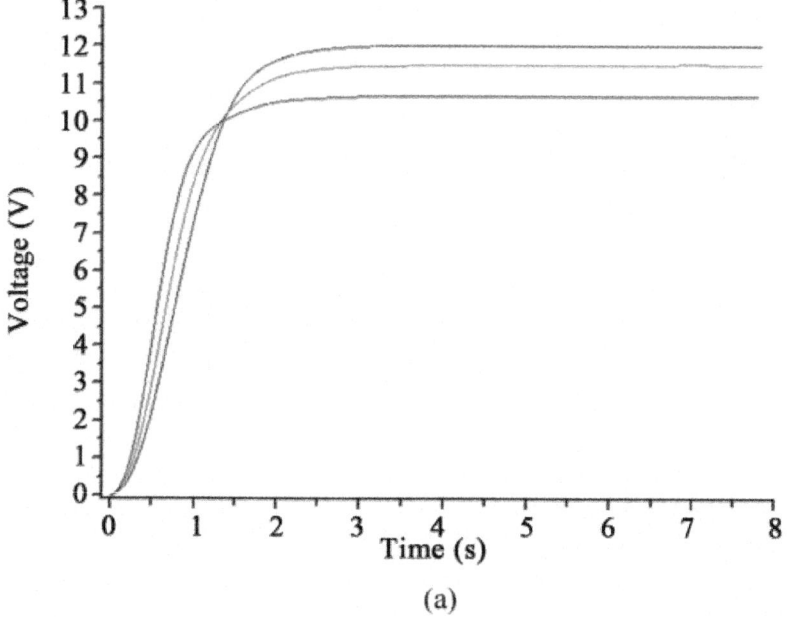

(a)

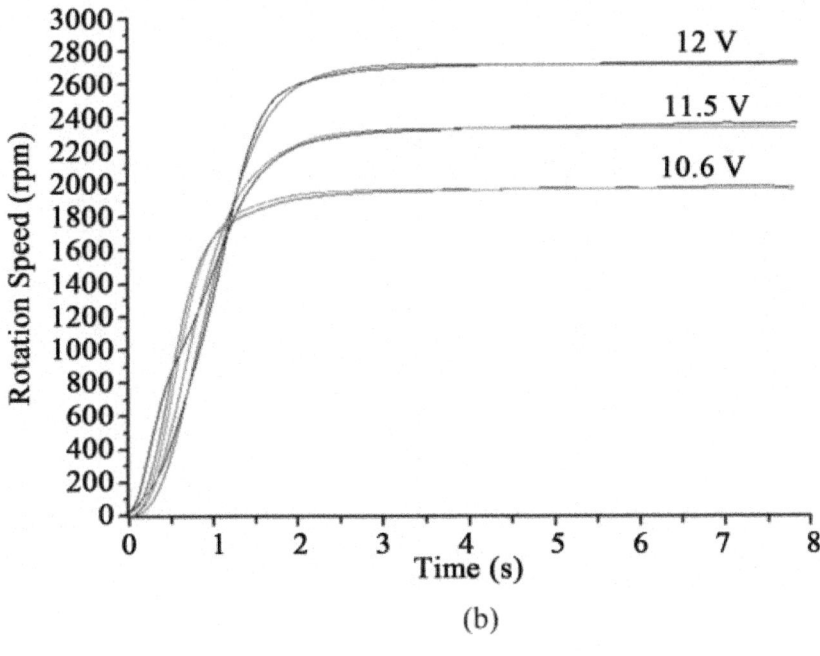

(b)

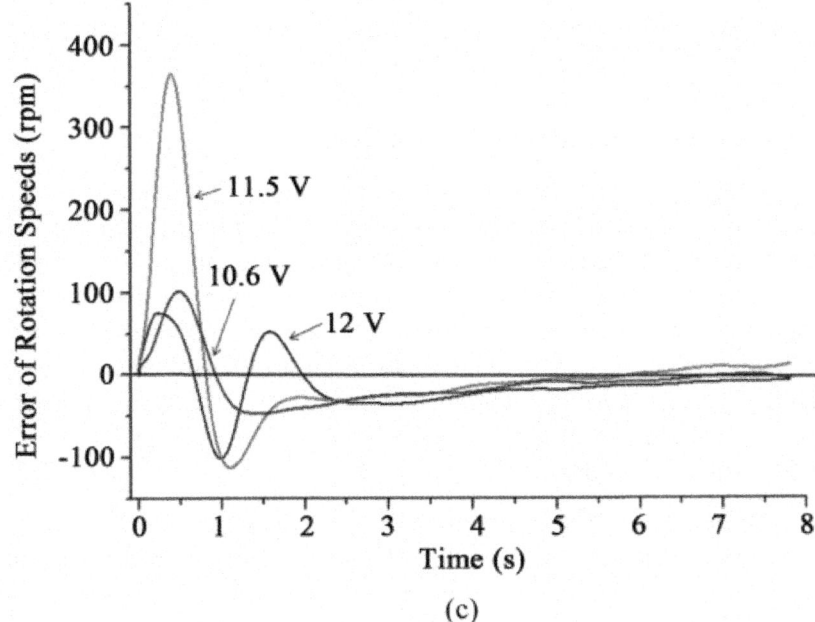

(c)

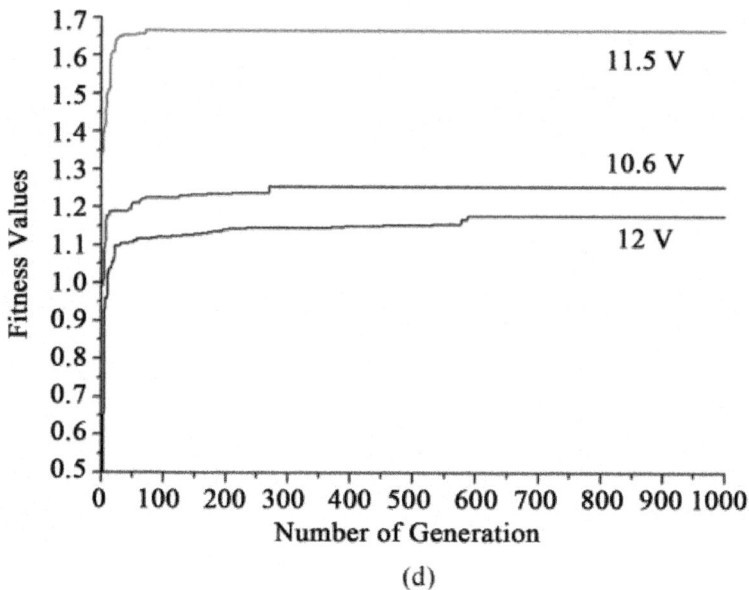

(d)

Figure 5: (a) The input voltages; (b) The comparisons in rotation speeds; (c) The errors of rotation speeds; (d) The convergences of the fitness values for three different input voltages.

Comparisons of Identified Parameters

The feasible domains and identified parameters by method1 of ARGA are shown in Table 3. For the three different input voltages, the identified parameters are little different, and the rotation speeds are analogous as show in Figure 5(b). It is noted that the wide feasible domains may affect the identified values of system parameters, and their settings need numerical experience for rapid convergence. It is concluded that the identified parameters will be better if the feasible domains are smaller.

For the Induction Motor

The electrical input voltages and rotation output speeds of a real induction motor are measured, and the ARGAs are employed to identify system's parameters in experiments. The experimental setup is shown in Figure 6, where the computer command is transformed by the driver to the induction motor. The input AC voltages and the rotation speeds are measured and transformed by the D/A card to numerical computation. The induction motor is three-phase with the rated specifications: 220 V, 60 Hz, 7 A, and 1720 rpm.

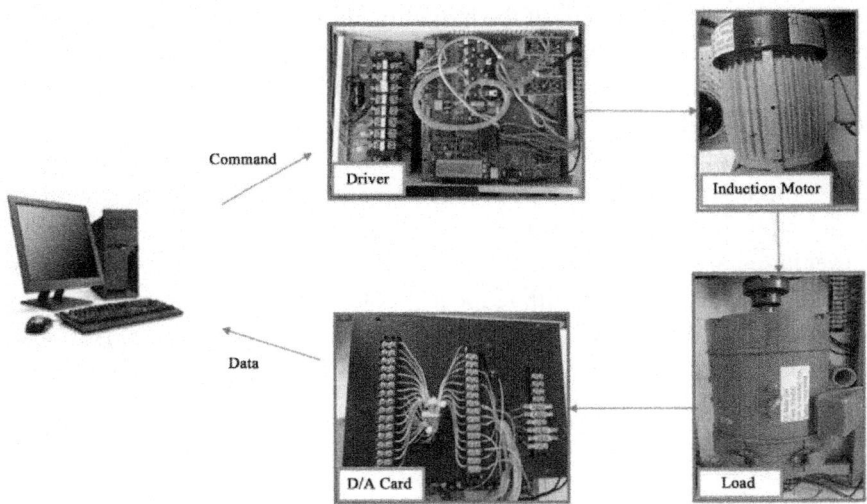

Figure 6: Experimental setup.

Table 3: The identified parameters of a BLDC motor for different input voltages

Parameters	Feasible domains	Identified values for a BLDC motor		
		10.6V	11.5V	12V
$R_a(\Omega)$	1 - 2	1.12	1.06	1.13
$L_a(H)$	4 - 5	4.06	4.03	4.65
$J_m \times 10^5 \, (\mathrm{N \cdot m \cdot s^2/rad})$	1 - 10	1.83	2.77	2.79
$B_m \times 10^5 \, (\mathrm{N \cdot m \cdot s/rad})$	1 - 10	1.67	1.07	1.58
$K_t \times 10^2 \, (\mathrm{N \cdot m/A})$	2.5 - 5.5	4.07	4.61	4.09
$\alpha \times 10^7 \, (\mathrm{m \cdot N \cdot m/rpm^2})$	0 - 2	0.00	1.14	0.50
Fitness Values		1.26	1.67	1.18

Comparisons between the Experimental and Identified Results

The input voltages are given to the induction motor for system's identification in experimental results. From numerical simulations, the method 1 of ARGA, which not only accurately search for parameters but also has small generation number in quick convergence, is the best method to identify parameters, and is also applied to the experimental identification.

In order to obtain experimental results, $V_0 = 90 \text{ V}$, $V_1 = 35 \text{ V}$, $\omega_0 = 5 \text{ rad/sec}$, $\omega_1 = 20 \text{ rad/sec}$, $T_1 = 3 \text{ sec}$ and $T_p = 15 \text{ sec}$ are taken in Equations (26) and (27) as the input AC voltages and shown in Figure 7(a). It is seen the rotation speeds of the induction motor obtained from LabVIEW are low before 10 sec, and the induction is unstable during this interval. Therefore, the identification is performed after 10 sec when the system is stable. Figure 7(b) compares the curves of the experimental and identified rotation speeds. It is seen that the identified rotation speeds are close to the experimental ones. The comparisons of the identified speeds by the SRGA and ARGA with respect to the experimental ones are shown in Figure 7(c). It is seen that the error percentages are very small, and the errors of the ARGA are smaller than the SRGA. The fitness value with respect to total number of generations is shown in Figure 7(d). The SRGA converges near the 200th generation. It is seen that the ARGA has faster convergence near the 50th generation and higher fitness value.

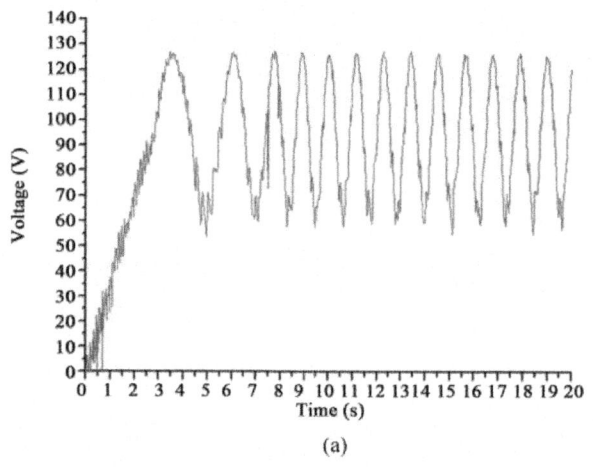

(a)

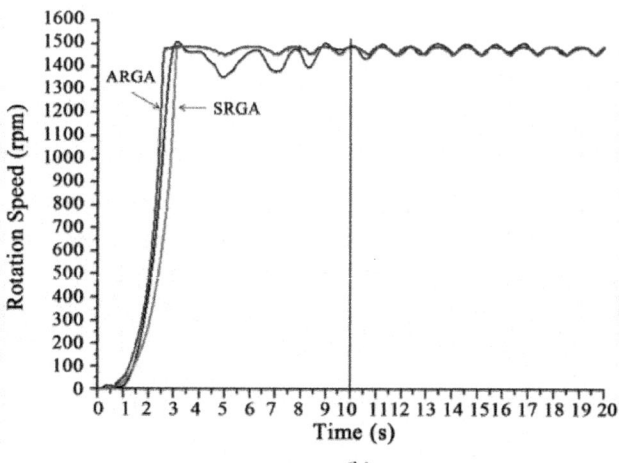

(b)

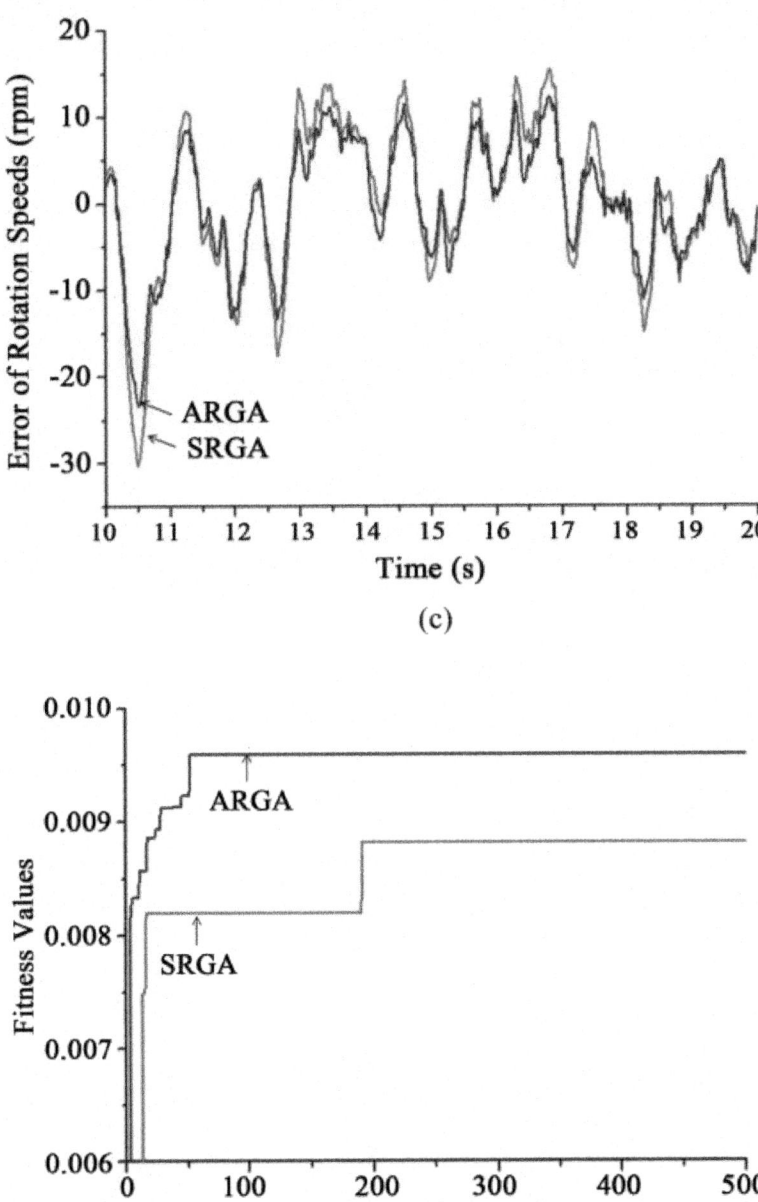

(c)

(d)

Figure 7: (a) The input voltages; (b) The comparisons in rotation speeds; (c) The errors of rotation speeds; (d) The convergences of the fitness values for the SRGA and ARGA.

In order to validate the identified parameters by the SRGA and ARGA in Equation (16), the input voltage as an exponential function in experiments is taken as follows:

$$v_m(t) = V \times \left(1 - e^{-\alpha t}\right)$$

(28)

where $\alpha = 5$ and V = 95, 150, 210 V, respectively. The input voltages $v_m(t)$ are shown in Figure 8(a). Substituting (28) into (25), the input electrical voltages $\left(v_{ds}, v_{qs}\right)$ are obtained, and then substituting the electrical voltages into (16), the rotation speeds are obtained. The rotation speeds for these three voltages are shown in Figures 8(b)-(d), respectively.

The electric currents $\left(i_{ds}, i_{qs}\right)$ can also be obtained from (16), and the input electrical current i_s can be obtained by $i_s = \left(i_{ds}^2 + i_{qs}^2\right)^{1/2}$ in Equation (19), and shown in Figures 9(a)-(c) for different voltages. It is seen that the electric currents by the SRGA and ARGA are similar, and the ARGA are better than SRGA.

Comparisons of Identified Parameters

The feasible domains and the identified parameters between the SRGA and method 1 of ARGA are compared in Table 4. It is seen the identified parameters are similar in experiments. However, the fitness value of the ARGA is bigger than the SRGA, and it means the ARGA parameters are more accurate and correct. For the convergence in generations, the ARGA converges at the 53thgeneration, and is faster than the SRGA.

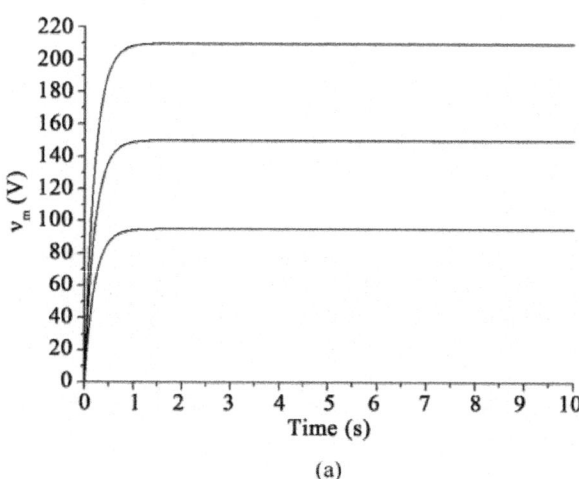

(a)

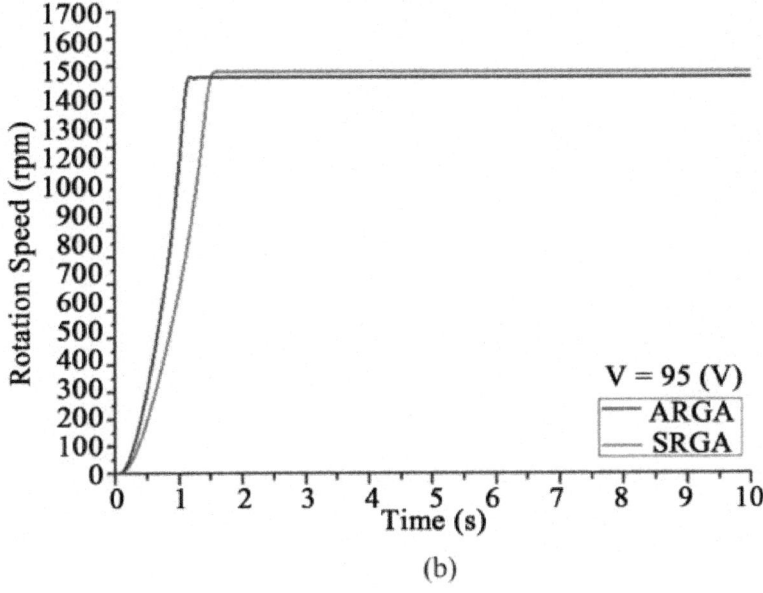

(b)

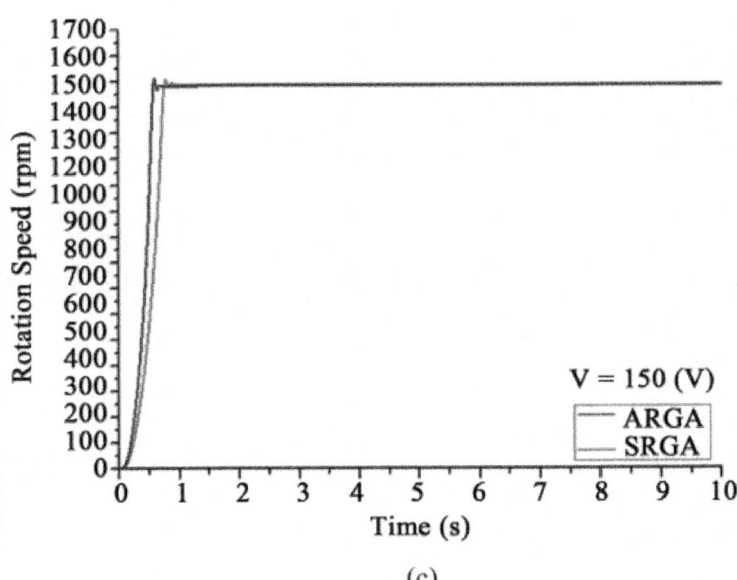

(c)

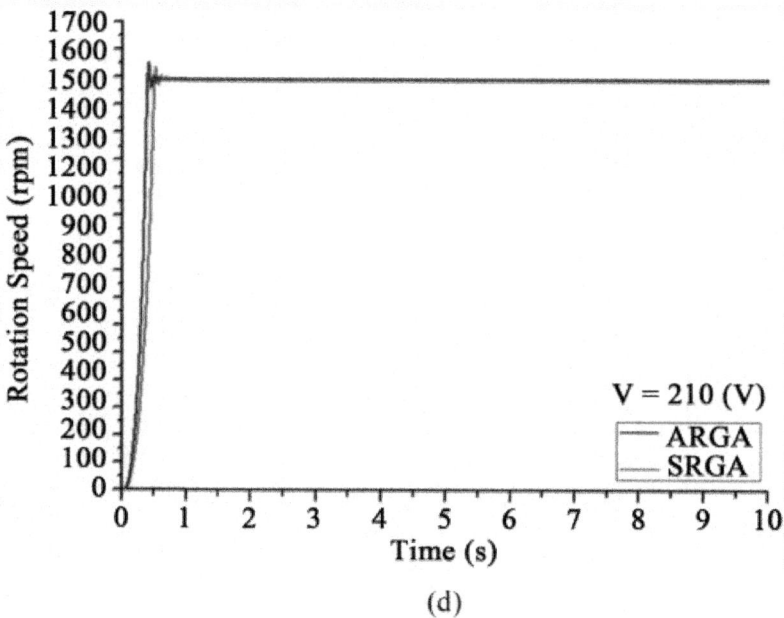

(d)

Figure 8: (a) The input voltages; (b) The rotation speeds for 95 V; (c) The rotation speeds for 150 V; (d) The rotation speeds for 210 V.

Table 4: The identified parameters for the SRGA and ARGAs

	Feasible domains	Identified values for an induction motor	
		SRGA	ARGA
$R_s(\Omega)$	0.0 - 1.5	0.507	0.489
$R_r(\Omega)$	0.0 - 1.5	0.689	0.720
$L_s(H)$	0.0 - 0.5	0.186	0.202
$L_r(H)$	0.0 - 0.5	0.302	0.276
$L_m(H)$	0.0 - 0.5	0.218	0.215
$J_m(N \cdot m \cdot s^2/rad)$	0.0 - 0.5	0.061	0.085
$B_m(N \cdot m \cdot s/rad)$	0.0 - 0.5	0.021	0.020
Fitness value	.	0.0088	0.0097
Convergence generation	.	191	53

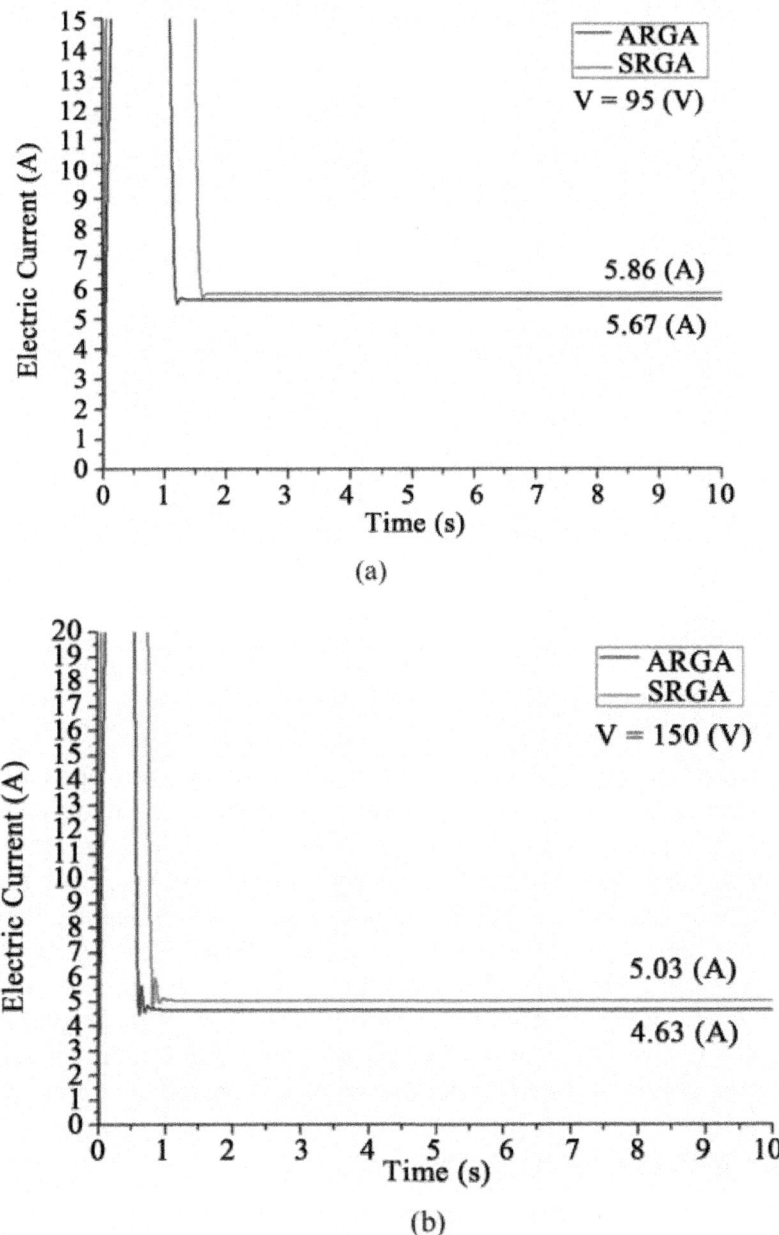

(a)

(b)

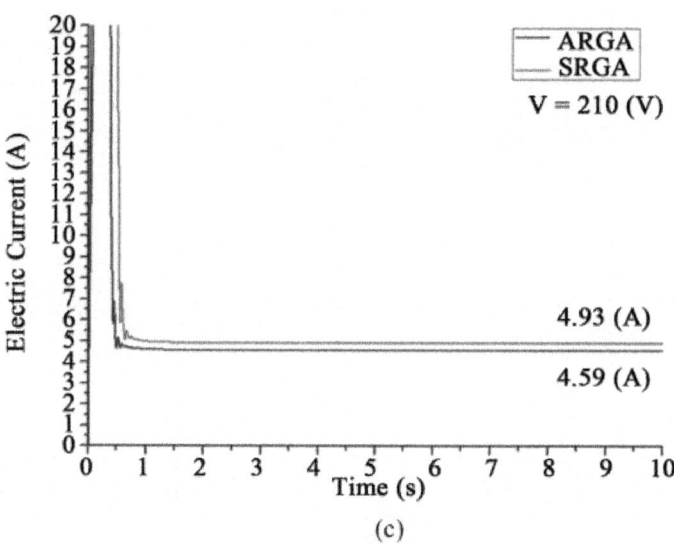

(c)

Figure 9. (a) The electric current for the input voltage 95 V; (b) The electric current for the input voltage 150 V; (c) The electric current for the input voltage 210 V.

CONCLUSION

This paper attempts to improve crossover and mutation operators in the traditional genetic algorithm by the adaptive technique. Effectiveness of the algorithm in identifying system's parameters is verified by the BLDC motor and induction motor. It is found that the SRGA and ARGA methods are feasible to system identification. The results show that the ARGA is found to have the faster convergence and the larger fitness value than the SRGA. In numerical simulations, the ARGAs with the identified parameters' errors percentages are less than 5% with respect to the assigned parameters. In this paper, method 1 of ARGA is found to be the best one to identify parameters of the BLDC and induction motors, and some experimental results are also compared.

ACKNOWLEDGEMENTS

The financial support from Ministry of Science and Technology of the Republic of China (MOST103-2221- E-327-009-MY3) is gratefully acknowledged.

REFERENCES

1.   Markovic, M., Hodder, A. and Perriard, Y. (2009) An Analytical Determination of the Torque-Speed and Efficiency-Speed Characteristics

of a BLDC Motor. IEEE Conversion Congress and Exposition, September 2009, 168-172.

2. Hemati, N. and Leu, M. (1990) A Complete Model Characterization of Brushless DC Motors. IEEE Transactions on Industry Applications, 28, 172-180. http://dx.doi.org/10.1109/28.120227

3. Lee, C.Y. and Hsieh, Y.H. (2012) Bearing Damage Detection of BLDC Motors Based on Current Envelope Analysis. Measurement Science Review, 12, 290-295. http://dx.doi.org/10.2478/v10048-012-0040-7

4. Fazil, M. and Rajagopal, K.R. (2011) Nonlinear Dynamic Modeling of a Single-Phase Permanent-Magnet Brushless DC Motor Using 2-D Static Finite-Element Results. IEEE Transactions on Magnetics, 47, 781-786. http://dx.doi.org/10.1109/TMAG.2010.2103955

5. Wang, S.J., Cheng, C.C., Lin, S.K., Ju, J.J. and Huang, D.R. (2005) An Automatic Pin Identification Method for a Three-Phase DC Brushless Motor. IEEE Transactions on Magnetics, 41, 3916-3918. http://dx.doi.org/10.1109/TMAG.2005.854970

6. Melkote, H. and Khorrami, F. (1999) Nonlinear Adaptive Control of Direct-Drive Brushless DC Motors and Applications to Robotic Manipulators. IEEE/ASME Transactions on Mechatronics, 4, 71-81. http://dx.doi.org/10.1109/3516.752086

7. Wang, W.J. and Wang, C.C. (1998) A New Composite Adaptive Speed Controller for Induction Motor Based on Feedback Linearization. IEEE Transactions on Energy Conversion, 13.

8. Krishnan, R. and Doran, F.C. (1987) Study of Parameter Sensitivity in High Performance Inverter Fed Induction Motor Drive System. IEEE Transactions on Industry Applications, IA-23, 623-635. http://dx.doi.org/10.1109/tia.1987.4504960

9. Toliyat, H.A., Levi, E. and Raina, M. (2003) A Review of RFO Induction Motor Parameter Estimation Techniques. IEEE Transactions on Energy Conversion, 18. http://dx.doi.org/10.1109/TEC.2003.811719

10. Wong, C.D., Wang, C.H. and Rung, R.F. (2013) System Identification of an Induction Motor. Chinese Society of Mechanical Engineers in Ilan University.

11. Srinivas, M. and Patnaik, L.M. (1994) Adaptive Probabilities of Crossover and Mutation in Genetic Algorithms. IEEE Transactions on System, Man, and Cybernetics, 24, 656-667. http://dx.doi.org/10.1109/21.286385

12. Ginley, B.M., Maher, J., Riordan, C.O. and Morgan, F. (2011) Maintaining Healthy Population Diversity Using Adaptive Crossover, Mutation, and

Selection. IEEE Transactions on Evolutionary Computation, 15, 692-714. http://dx.doi.org/10.1109/TEVC.2010.2046173

13. Jiang, Y., Jiang, J. and Zhang, Y. (2011) A Novel Fuzzy Multi-Objective Model Using Adaptive Genetic Algorithm Based on Cloud Theory for Service Restoration of Shipboard Power Systems. IEEE Transactions on Power Systems, 27, 612-620. http://dx.doi.org/10.1109/TPWRS.2011.2179951

14. Schaible, U. and Szabados, B. (1999) Dynamic Motor Parameter Identification for High Speed Flux Weakening Operation of Brushless Permanent Magnet Synchronous Machines. IEEE Transactions on Energy Conversion, 14, 486-492. http://dx.doi.org/10.1109/60.790901

15. Goldberg, D.E. and Holland, J.H. (1988) Genetic Algorithms and Machine Learning. Machine Learning, 3, 95-99.

16. Ha, J.L., Fung, R.F. and Han, C.F. (2005) Optimization of an Impact Drive Mechanism Based on Real-Coded Genetic Algorithm. Sensors and Actuators A: Physical, 121, 488-493. http://dx.doi.org/10.1016/j.sna.2005.03.001

17. Chang, P.C., Hsieh, J.C. and Wang, C.Y. (2007) Adaptive Multi-Objective Genetic Algorithms for Scheduling of Drilling Operation in Printed Circuit Board Industry. Applied Soft Computing, 7, 800-806. http://dx.doi.org/10.1016/j.asoc.2006.02.002

18. Raman, S. and Patnaik, L.M. (1996) Performance-Driven MCM Partitioning through an Adaptive Genetic Algorithm. IEEE Transactions on Very Large Scale Integration Systems, 4, 434-444. http://dx.doi.org/10.1109/92.544408

19. Wang, F., Li, J., Liu, S., Zhao, X., Zhang, D. and Tian, Y. (2014) An Improved Adaptive Genetic Algorithm for Image Segmentation and Vision Alignment Used in Microelectronic Bonding. IEEE/ASME Transactions on Mechatronics, 19, 291-296. http://dx.doi.org/10.1109/TMECH.2013.2260555

20. Wang, L. and Tang, D.B. (2011) An Improved Adaptive Genetic Algorithm Based on Hormone Modulation Mechanism for Job-Shop Scheduling Problem. ELSEVIER Expert Systems with Applications, 38, 7243-7250. http://dx.doi.org/10.1016/j.eswa.2010.12.027

21. Yuan, L.Z., Xiao, B. and Wei, X.J. (2010) BP Network Model Optimized by Adaptive Genetic Algorithms and the Application on Quality Evaluation for Class Teaching. IEEE International Conference on Future Computer and Communication, 3, 273-276. http://dx.doi.org/10.1109/icfcc.2010.5497635

22. Du, Y., Fang, J. and Miao, C. (2014) Frequency-Domain System Identification of an Unmanned Helicopter Based on an Adaptive Genetic Algorithm. IEEE Transactions on Industrial Electronics, 61, 870-881. http://dx.doi.org/10.1109/TIE.2013.2257135

23. Yang, X. (2012) Quantitative Detection for Gas Mixtures Based on the Adaptive Genetic Algorithm and BP Network. Proceedings of the IEEE International Conference on Industrial Control and Electronics Engineering, Xi'an, 23-25 August 2012, 1341-1344.

24. Lin, G. and Liu, G. (2010) Tuning PID Controller Using Adaptive Genetic Algorithms. Proceedings of the IEEE International Conference on Computer Science and Education, Hefei, 24-27 August 2010, 519-523. http://dx.doi.org/10.1109/iccse.2010.5593559

25. Chang, C.Y. and Chen, D.R. (2010) Active Noise Cancellation without Secondary Path Identification by Using an Adaptive Genetic Algorithm. IEEE Transactions on Instrumentation and Measurement, 59, 2315-2327. http://dx.doi.org/10.1109/TIM.2009.2036410

26. Xu, X.Q. and Lei, L. (2011) The Research of Advances in Adaptive Genetic Algorithm. Proceedings of the IEEE International Conference on Signal Processing, Communications and Computing (ICSPCC), Xi'an, 14-16 September 2011, 1-6.

27. Lin, C. (2009) An Adaptive Genetic Algorithm based on Population Diversity strategy. Proceedings of the 3rd International Conference on Genetic and Evolutionary Computing, Guilin, 14-17 October 2009, 93-96. http://dx.doi.org/10.1109/wgec.2009.67

28. Potts, J.C. and Giddens, T.D. (1994) The Development and Evaluation of an Improved Genetic Algorithm Based on Migration and Artificial Selection. IEEE Transactions on Systems, Man, and Cybernetics, 24, 73-86. http://dx.doi.org/10.1109/21.259687

29. Li, T.H., Lucasius, C.B. and Katerman, G. (1992) Optimization of Calibration Data with the Dynamic Genetic Algorithm. Analytica Chimica Acta, 268, 123-134. http://dx.doi.org/10.1016/0003-2670(92)85255-5

30. Li, J.H. and Li, M. (2008) Genetic Algorithm with Dual Species. Proceedings of the IEEE International Conference on Automation and Logistics, Qingdao, 1-3 September 2008, 2572-2575. http://dx.doi.org/10.1109/ical.2008.4636604

31. Nowostawski, M. and Poli, R. (1999) Parallel Genetic Algorithm Taxonomy. Proceedings of the IEEE International Conference on Knowledge-Based Intelligent Information Engineering Systems,

Adelaide, 31 August-1 September 1999, 88-92. http://dx.doi.org/10.1109/kes.1999.820127

32. Alba, E. and Tomassini, M. (2002) Parallelism and Evolutionary Algorithms. IEEE Transactions on Evolutionary Computation, 6, 443-462. http://dx.doi.org/10.1109/TEVC.2002.800880

33. Chen, F., Jiang, B., Tao, G. and Zhang, X. (2013) Improved Adaptive Genetic Algorithm For Grid Resource Scheduling via Quantum Control Techniques. Proceedings of the IEEE Control and Decision Conference, Guiyang, 25-27 May 2013, 3974-3978. http://dx.doi.org/10.1109/ccdc.2013.6561644

34. Jiang, Y., Jiang, J. and Zhang, Y. (2012) A Novel Fuzzy Multi-Objective Model Using Adaptive Genetic Algorithm Based on Cloud Theory for Service Restoration of Shipboard Power Systems. IEEE Transactions on Power Systems, 27, 612-620. http://dx.doi.org/10.1109/TPWRS.2011.2179951

35. Shao, J., Nolan, D., Teissier, M. and Swanson, D. (2003) A Novel Microcontroller-Based Sensorless Brushless DC (BLDC) Motor Drive for Automotive Fuel Pumps. IEEE Transactions on Industry Applications, 39, 1734-1740.

36. Kuria, J. and Hwang, P. (2011) Investigation of Thermal Performance of Electric Vehicle BLDC Motor. International Journal of Mechanical Engineering, 1, 1-17.

37. Liaw, C.M., Chao, K.H. and Lin, F.J. (1992) A Discrete Adaptive Field-Oriented Induction Motor Drive. IEEE Transactions on Power Electronics, 7, 411-419.

38. Attaianese, C., Damiano, A., Gatto, G., Marongiu, I. and Perfetto, A. (1998) Induction Motor Drive Parameters Identification. Transactions on Power Electronics, 13.

39. Wang, W.J. and Wang, C.C. (1999) Speed and Efficiency Control of an Induction Motor with Input-Output Linearization. IEEE Transactions on Energy Conversion, 14, 373-378.

40. Shaw, S.R. and Leeb, S.B. (1999) Identification of Induction Motor Parameters from Transient Stator Current Measurements. IEEE Transactions on Industrial Electronics, 46, 139-149. http://dx.doi.org/10.1109/41.744405

41. Zaky, M.S., Khater, M.M., Shokralla, S.S. and Yasin, H.A. (2009) Wide-Speed-Range Estimation with Online Parameter Identification Schemes of Sensorless Induction Motor Drives. IEEE Transactions on Industrial Electronics, 56, 1699-1707. http://dx.doi.org/10.1109/TIE.2008.2009519

42. Liaw, C.M., Kung, Y.S. and Wu, C.M. (1991) Design and Implementation of a High-Performance Field-Oriented Induction Motor Drive. IEEE Transactions on Industrial Electronics, 38, 275-282.

Chapter 5

WHIRL INTERACTION OF A DRILL BIT WITH THE BORE-HOLE BOTTOM

Nabil W. Musa[1], V. I. Gulyayev[2], L. V. Shevchuk[2], and Hasan Aldabas[1]

[1]Department of Mechanical Engineering, Philadelphia University, Amman, Jordan

[2]Department of Mathematics, National Transport University, Kyiv, Ukraine

ABSTRACT

This paper deals with the theoretic simulation of a drill bit whirling under conditions of its contact interaction with the bore-hole bottom rock plane. The bit is considered to be an absolutely rigid ellipsoidal body with uneven surface. It is attached to the lower end of a rotating elastic drill string. In the perturbed state, the bit can roll without sliding on the bore-hole bottom, performing whirling vibrations (the model of dynamic equilibrium with pure rolling when maximum cohesive force does not exceed the ultimate Coulombic friction). To describe these motions, a nonholonomic dynamic model is proposed, constitutive partial differential equations are deduced. With their use, the whirling vibrations of oblong and oblate ellipsoidal bits are analyzed, the functions of cohesive (frictional) forces are calculated. It is shown that the system of elastic drill string and ellipsoidal bit can acquire stable or unstable whirl modes with approaching critical Eulerian values by the parameters of axial force, torque and angular velocity. The analogy of the found modes of motions with ones of the Celtic stones is established. It is shown that the ellipsoidal bits can stop their whirling vibrations and change directions of their circumferential motions in the same manner as the ellipsoidal Celtic stones do. As this takes place, the trajectories of the oblate ellipsoidal bits are characterized by more complicated paths and irregularities.

INTRODUCTION

Late in the 19th century, one of the most fascinating dynamic effects of the nonholonomic mechanics was found. It is associated with the so called problem of Celtic stones consisting in the apparent violation of the physical law of angular momentum conservation. This phenomenon is realized for solid ellipsoidal bodies with some breakdowns of geometry or inertia symmetries (the Celtic stones). If to put this body into touch with an uneven horizontal plane and to twirl it around the vertical axis, then, after a time, it ceases to rotate and after some short-run irregular vibration relative to a horizontal axis ("dances"), it again begins to rotate around the vertical axis, yet this time, in the opposite direction.

The similar changes of the rotation directions occur repeatedly for some shapes of bodies. In so doing, the point of the body contact with plane traces out a rather complicated paths. At the stages of rotary motions, they have the shapes of expanding or narrowing spirals, while during the transition through the vibration regimes they acquire the outlines of complicated curves with loops and cuspidal points.

The curious stones possessing these properties were originally called celts because their behavior was discovered by archaeologists studying the prehistoric utensils of ancient Celtic folks. The first description and physical explanation of their motion was presented by G.T. Walker in 1895 [1] . Subsequently, two basic dynamic models were used for analysis of queer behavior of the Celtic stones. The more general and complex statement of the problem consists in investigation of the body motion on the horizontal plane with allowance made for the sliding effect and presence of a friction force at the contact point between the bodies. By virtue of its severity, it turned out to be less attractive and fruitful. The nonholonomic model of the Celtic stone motion was found to be simpler and more obvious, so with its aid, the principal properties and qualitative features of the body motion were revealed [2] - [4] .

The outwardly similar motions can be performed by the polycrystalline-diamond-compact (PDC) bit of a deep drill string (DS). As the bit is connected to the lower end of the DS, their motions are interdependent. In drilling, the DS thrusts the bit against the bore-hole bottom and by-turn is subjected to action of the vertical reaction force, torque and centrifugal inertia force, which conduce to decrease its bending stiffness and lead to its elastic bending. As noted in [5] - [7] , in consequence of its deforming, the coaxiality of the DS and bit is disaligned, the bit shifts out of the system center and begins to roll on the bore-hole bottom.

In this case, two motion regimes are possible which are differed by the character of the bit interaction with the bore-hole bottom and, consequently, are described by different mathematic models. Thus, if the force of the bit thrust to the bottom is not large and adhesion between them is disrupted, then the typical frictional interaction described by the Coulomb friction law prevails. Analysis of the bit dynamics is performed with the use of such model in paper [8] .

But the situation changes essentially with enlargement of the compressive thrust force and its approaching to the Eulerian critical value. Then, the bending stiffness of the DS reduces, it hogs and the diamond impregnations existing in the bit surface penetrate into the rock medium. As this takes place, the bit axis tilts and nutates. It should be particularly emphasized that at this stage, the bit is under action of the vertical force (weight on bit) which essentially exceeds all other forces. It presses the bit to the hole bottom (not to the hole wall) and the bit loses its ability to slide on the bore-hole bottom and begins to roll on its surface, lagging behind or outstripping the DS rotation. As a result, the point of the bit contact with the reference surface (the instantaneous centre of velocities) may describe extremely complicated trajectories with loops and cuspidal points and to change the motion directions, as happens with the Celtic stones.

Analyses of such systems motions were performed, for the most part, with the use of natural models or with the help of simplified mathematic models. In early study of 1990, Brett et al. [9] assumed with the use of simple mass-spring model that the main cause of PDC bit failures was their backward whirl, representing a specific kind of its lateral vibration. That is, a bit may vibrate laterally but may not necessarily be whirling. Only when the bit's geometric center moves around the hole's center line is the bit valid whirling. During whirl, the instantaneous center of rotation moves around the bit, and it whirls backward around the hole. Cutters on a whirling bit can move sideways, backward, and much faster than those on a true rotating bit. Laboratory and field results showed the detrimental effects of whirl on the bit rate of penetration and its life. Special attention was also paid to mechanisms of cutting and moving of bits with sharp and dull cutters. In 1992 Langeveld [10] elaborated fully dynamic and three-dimesional model of torsional, axial and lateral vibrations of different PDC bit designs. His analyses demonstrated that conventional PDC bits tend to whirl backward, while anti-whirl PDC bits effectively reduce this tendency and whirl forward. The possibility to vibrate with backward and forward whirlings was noted also by Schen et al. [11] and Thor Viggo and Age [12] . In their observations, the whirl rotation occurred several times per revolution of the bit and during whirl, the bit drilled an overgage hole with a multilobed cross-section. Chen et al. [13] accentuated that in most cases bit whirl (backward or forward) is coupled with other forms of lateral vibrations, therefore, the trajectory of the bit center is usually not a circle. Wu et al. [14] and Ledgerwood et al. [15] [16] bind initiation of whirling vibrations with torsional stick/slip friction vibrations and consider it to be the primary cause of bit damage. Sowers et al. [17] maintain this point of view and assert that additional detrimental effects caused by coupling of these phenomena include the development of ledges, bore-hole oscillations and spiraling, and premature failure of downhole tools. They consider that bore-hole quality can be improved via the use of roller reamers. Johnson [18] points out that bit whirl is responsible for bit-rock interaction, leading to lateral vibrations, and results in short runs, low rate of penetration, high cost per foot, poor hole quality, and downhole-tool damage. The methods developed for the bit whirl reduction is recorded to mitigate the detrimental effects, but do not completely eliminate them and there are compromises associated with each method that prevent their usage in many applications. Friction force influence on the bit whirling is discussed by Christoforou and Yigit [6] and Leine et al. [7] . With the use of low degree of freedom models, simulating the bit rolling on the well wall surface, they provide qualitative insight into rolling with sliding and pure rolling. As shown by calculations, for the time duration covered by simulation, the initial forward whirl changes into a transitory phase, in which the whirl direction continuously changes between forward and backward, and eventually settles into a backward whirl. Note once more, that the similar motions are performed by Celtic stones [1] - [4] . The difference between sliding and rolling motions is also emphasized by Stroud et al. [19] : "Backward whirl occurs when the drill string switches from being in sliding contact to rolling contact with the borehole. Once a drillstring assembly goes into rolling contact with the borehole, it rotates backwards around the center point". Kovalyshen [20] elaborated a simple analytical model of bit whirl, taking into account the bit geometry characterized by three-dimensional parameters. Depending on their values, the analysis results have shown that the system can be stable or undergo forward or backward whirl. Spanos et al. [21] and Ritto et al. [22] underline that the uncertainty is inherent in the properties of parameters determining the drilling processes, therefore, the stochastic nature of the bit rock interaction should be taken into consideration. General questions of friction influence on deep hole drilling processes are outlined by R. Samuel [23] .

This paper is dedicated to analysis of pure rolling of the bit. Inasmuch as the constraints imposed on the system under such conditions are kinematic, it is worthwhile to apply the methods of nonholonomic mechanics for investigation of its dynamic motion. In paper [24] , on the basis of such statement of the problem, dynamics of spinning and rolling without sliding of spherical bits on spherical surfaces of bore-hole bottoms is studied. It is noted, that the bit whirl vibration can be entailed by three kinds of stable and unstable motions associated with forward and inverse rolling of the bit and its pure spinning.

As this takes place, the modes of these motions essentially depend on the DS flexibility which is determined not only by its mechanical stiffness, but also by proximity of its state to the Eulerian instability. The inference about the necessity to consider the motion of ellipsoidal bits is drawn.

However, as shown in [25] , the problem of rolling a body on an uneven plane becomes more complicated with transfer from spherical bodies to ellipsoidal ones. Nevertheless, as the bits in the shapes of elongated and oblate ellipsoids are widely used in the practice of deep drilling, the question of how their shape geometry influences on the processes of their whirling vibrations is of great interest. This question is considered below.

THREE-POINT BOUNDARY VALUE PROBLEM FOR ELAS-TIC VIBRATIONS OF THE LOWER SECTIONS OF A DRILL STRING

To study the phenomenon of a bit rolling without sliding on a bore-hole bottom surface, it is necessary to state the problem about elastic bending vibration of the DS taking into account the constraints fulfilling the roles of boundary conditions for the dynamic equations of the DS.

Consider that it rotates with constant angular velocity ω. At its upper part, the DS is prestressed by tensile gravity force, its lower part is compressed by the contact reaction force reducing bending stiffness of the system. To enlarge this stiffness, usually centralizing devices are located in the lower segments of the DS which serve as supplemented supports. Because the most intensive bending vibrations of a DS prevail in the sections nearest to the drill bit, occurrence of the upper sections will not be taken in account and only two lower sections (segments AB and BC in Figure 1) are separated for the simulation.

Figure 1: The scheme of the drill sections separated for analysis.

Vibrations of these sections are analyzed on the basis of the theory of compressed-twisted rotating beams with internal flows of washing liquid (mud). Phenomena of torsion vibrations of rotating DSs under conditions of friction interaction between its bit and destructed rock is studied in papers [26] - [28] . Influence of internal flows of liquid on the tube rod dynamics is analyzed in [29] [30] . It is shown that these flows further the system instability.

To study elastic bending dynamics of the DS under consideration, introduce inertial coordinate system OXYZ with its origin O at the end A and axis OZ in line with the DS axis, as well as the Oxyz coordinate system rotating together with the DS tube. Let i, j, k be the unit vectors of this system.

It is assumed that the bending stress-strain state of the DS is determined by the lateral displacements $u(z)$, $v(z)$ in the planes xOz, yOz of the Oxyz system.

In the deduction of differential equations of the DS dynamics, take into account that it is prestressed by internal axial force T and torque M_z. Besides, it rotates with angular velocity ω and washing liquid moves with linear velocity v in its cavity. As a consequence, in [31] [32] , the following form of the constitutive equations was constructed.

$$EI\frac{\partial^4 u}{\partial z^4} - T\frac{\partial^2 u}{\partial z^2} - M_z\frac{\partial^3 v}{\partial z^3} - \gamma_t\omega^2 u - 2\gamma_t\omega\frac{\partial v}{\partial t} + V^2\gamma_t\frac{\partial^2 u}{\partial z^2} + 2V\gamma_t\frac{\partial^2 u}{\partial z\partial t} + \gamma_t\frac{\partial^2 u}{\partial t^2} = 0,$$

$$EI\frac{\partial^4 v}{\partial z^4} - T\frac{\partial^2 v}{\partial z^2} + M_z\frac{\partial^3 u}{\partial z^3} - \gamma_t\omega^2 v + 2\gamma_t\omega\frac{\partial u}{\partial t} + V^2\gamma_t\frac{\partial^2 v}{\partial z^2} + 2V\gamma_t\frac{\partial^2 v}{\partial z\partial t} + \gamma_t\frac{\partial^2 v}{\partial t^2} = 0. \quad (1)$$

Here EI is the bending stiffness of the DS: γ_l, γ_t are the linear masses of the washing liquid flow and drill string tube with internal liquid, respectively; v is the velocity of the washing liquid and t is the time.

The linear masses γ_l, γ_t are calculated through the equalities

$$\gamma_l = \rho_l f_l, \quad \gamma_t = \rho_l f_l + \rho f, \quad (2)$$

where ρ_l, ρ are the densities of the liquid and tube material; f_l, f are the cross-section areas of the internal channel of the DS tube and its wall, respectively.

In Equation (1), a dominant role is played by the terms with multipliers T and M_z, because they influence on the Eulerian stability of the DS section and thereby determine its bending stiffness. The first ones give rise to its buckling, while the second ones lead to spiral modes of the DS deforming [32] .

The roles of the members containing multipliers ω^2 and ω are less significant, because the influence of the DS rotating is essential in its long spans. Of even less importance are the terms containing the multipliers V^2 and v [31] [32] , nevertheless they are not omitted in the analysis.

Equation (1) together with appropriate boundary equations at points A, B, C (Figure 1) represent the three- point boundary value problem relative to the independent variable z. To analyze whirling vibrations of the system, some initial perturbations will be imparted to it and its response to them will be studied. This part of analysis is performed on the basis of the Cauchy problem statement.

STATEMENT OF BOUNDARY CONDITIONS FOR THE CONSTITUTIVE EQUATIONS

Peculiar features of the considered problem consist in its boundary conditions. At the points A and B, they are easily formulated. At the A point positioned between the conditionally discarded upper section and one chosen for analysis, the equations

$$u = v = 0, \quad \frac{\partial^2 u}{\partial z^2} = \frac{\partial^2 v}{\partial z^2} = 0$$

(3)

can be assumed. They correspond to the conditions of inverse symmetry of the modes of the Eulerian stability loss and free vibrations of a multispan beam.

At the B point, the equations are valid. They stem from conditions of continuity of elastic displacement μ and v.

$$u = v = 0, \quad \frac{\partial u}{\partial z}\bigg|_{z_B-0} = \frac{\partial u}{\partial z}\bigg|_{z_B+0}, \quad \frac{\partial v}{\partial z}\bigg|_{z_B-0} = \frac{\partial v}{\partial z}\bigg|_{z_B+0},$$

$$\frac{\partial^2 u}{\partial z^2}\bigg|_{z_B-0} = \frac{\partial^2 u}{\partial z^2}\bigg|_{z_B+0}, \quad \frac{\partial^2 v}{\partial z^2}\bigg|_{z_B-0} = \frac{\partial^2 v}{\partial z^2}\bigg|_{z_B+0},$$

(4)

The boundary equations at the C point are considerably more complicated and because of this, they should be especially considered on the basis of the bit motion model.

The problem of this model creation is associated with essential theoretical and technical obstacles because many different vibration phenomena occur simultaneously in the bit dynamics, doing it difficult to isolate, evaluate, and explain any one of them. To varying degrees, axial, torsional, and bending vibrations are all present and coupled. Bit bounce, stick-slip, forward and backward whirl and linear and parametric interconnection between axial and bending vibrations all occur. Moreover, these phenomena depend on many structural, mechanical and technical parameters. Therefore, the objective of our work is to improve the understanding of the interdependence between whirling, contact cohesion of the bit with hole bottom, the bit geometry, and elastic bending pliability of the DS. However, for a real drill string configuration, the simulation results should only be interpreted in a qualitative sense.

In our work, the initial stage of the whirl proceeding is considered. Therefore, the following assumptions are used.

1. Whirl vibrations of a PDC bit is analysed. It is considered as a rigid oblong or oblate ellipsoid.

2. The bit is connected with lower sections of elastic DS. Because of this, the displacements and nutations of the bit are provoked by elastic bending and tilting of the DS.

3   Initial stage of the whirling process is considered when the point of contact of the bit with rock abandoned the hole bottom centre and moves around it, but does not achieve the hole wall. The lateral displacements of the DS are small and it does not also touch the hole wall.

The vertical force (weight-on-bit) compresses the bit to the hole bottom and thus, they are in permanent contact.

1. Angular velocity ω of the DS rotation is considered to be constant.

2. The contact interaction of the bit with the hole bottom rock is described by the Coulombic friction law [23] [25] . According to it, the friction force F^{fr} is determined by the equality

$$F^{fr} \leq \mu N. \tag{5}$$

Here μ is the friction coefficient, N is the normal force of contact interaction.

Graphically, this law is represented in Figure 2, where v^{sl} is the relative sliding velocity of contacting bodies, $F^{ult} = \mu N$ is the ultimate friction force.

Then, if the coercive forces, acting on the bit in tangential direction, are less the ultimate friction force F^{ult}, the bit sliding is impossible and it rolls on the bottom surface. In this case the friction force is less than the value μN, it is unknown and can be determined by dynamic equations. This force is located in vertical segment of the diagram in Figure 2 and is named as the force of static friction or cohesive force.

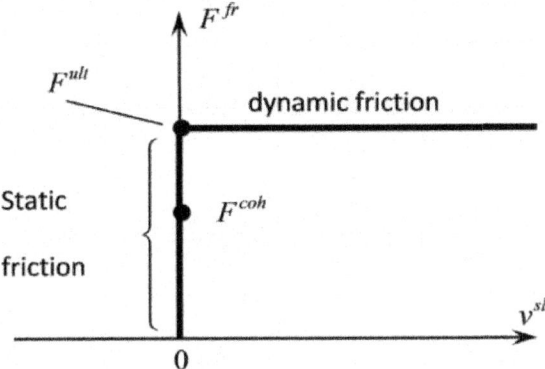

Figure 2: Schematic of friction force F^{fr} change.

When the coercive forces exceed the ultimate friction force F^{ult}, the relative accelerating motion of contacting bodies occur with constant friction force $F^{fr} = \mu N$. This regime is represented by horizontal segment in Figure 2 and is called dynamic friction. The first type of motion can be concluded to be typical for bits with sharp cutters, while the second regime is inherent to worn out bits.

In the proposed model, the pure rolling mechanism [7] of the bit whirling is assumed and the motion of its instantaneous centre of velocities around the hole axis [19] is analysed. The model gives the possibility to simulate the change of forward and backward whirl by each other and to explain essential enlargement of the whirl velocity in comparison with the DS rotation velocity.

Having these assumptions, it became possible to formulate equations of the body motion under action of elastic forces and moments generated at the DS end owing to contact interaction between the body and the surface.

In paper [24], the problem on rolling of a spherical bit on a curvilinear surface is considered. As further researches of the authors demonstrated, the attempts to generalize this problem to the case of the ellipsoidal bit rolling on a curvilinear surface cause severe difficulties. For this reason, below the problem on the rotary ellipsoid bit whirling on the plane uneven surface π (Figure 1) is considered.

It is pertinent to note that the problem on ellipsoid body rolling on a plane surface has a long-standing history in the nonholonomic mechanics [25] and yet, as a rule, only its simplest versions concerning free bodies under action of gravity forces were considered and only some partial solutions were constructed. A similar situation occurs also with respect to the Celtic stones problem [1] - [4].

In our case, the ellipsoidal bit spins relative to the axis of its symmetry and is subjected to action of the axial force t and elastic forces and moments. Besides, the equations of the bit motion do not make up constitutive equations, as in the problem on Celtic stones, but are no more than boundary conditions for the equations of the DS vibrations.

To deduce these equations detach conditionally the bit from the DS at the c point. Assume that the onset of whirling vibration is at the initial state, the bit can move in a narrow clearance without contact with the bore-hole side surface, and the bore-hole bottom is plane.

The bit motion is described by two vector equations. The first one represents the condition of kinematic compatibility of the motion velocities of the DS end and the bit body at the point of their connection. It represents the equation of nonholonomic constraint. The second equation is formulated reasoning from the condition of dynamic (in the simplified case, quasi-static) equilibrium of the elastic shear forces and moments applied to the bit.

To formulate the equations of the bit rolling, rigidly fix to it the $Cx_1y_1z_1$ coordinate system. In the undeformed state, axes Cx_1, Cy_1 are parallel to corresponding axes of the Oxyz system. Assume that under condition of elastic bending, the angles of the $Cx_1y_1z_1$ system slewing relative to the Oxyz system are small (Figure 3), so it is possible to introduce the vector of the slewing angle

$$\theta = -v'i + u'j + 0 \cdot k = \theta_x i + \theta_y j.$$
(6)

In the deformed state, the Cz_1 axis slews through the θ angle in the σ plane going through ellipsoid apex D, its center C, and the G point of contact of the bit with the π plane (Figure 4).

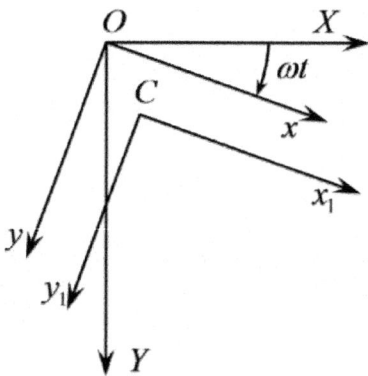

Figure 3: Top view of the used coordinate systems.

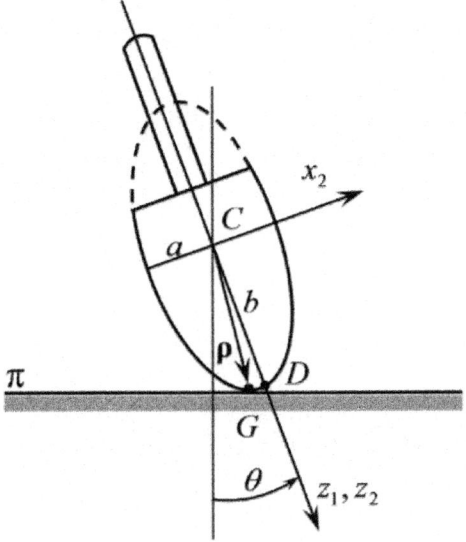

Figure 4: The position of touch point G in the plane of ellipsoid inclination.

For the purpose of determination of the G point in the Oxyz system, introduce the right-hand coordinate system $Cx_2y_2z_2$ with the Cz_2 axis running along the Cz_1 axis and the Cx_2 axis lying in the σ plane (Figure 4).

Then, it is possible to find the θ angle with the use of the formula

$$\theta = \pm\sqrt{(u')^2 + (v')^2} \tag{7}$$

and thereafter, to calculate angle α between the planes σ and xOz:

$$\alpha = -arctg\left(\theta_x/\theta_y\right). \tag{8}$$

With the aid of the introduced angle, one can find the vector ρ connecting the points C and G (Figure 4):

$$\rho = \frac{(b^2 - a^2)\sin\theta_y \cos\theta_y}{\sqrt{a^2 \sin^2\theta_y + b^2 \cos^2\theta_y}} i - \frac{(b^2 - a^2)\sin\theta_x \cos\theta_x}{\sqrt{a^2 \sin^2\theta_x + b^2 \cos^2\theta_x}} j + \sqrt{a^2 \sin^2\theta + b^2 \cos^2\theta}\, k. \tag{9}$$

Here a and b are the semi-axes of the bit ellipsoid.

At the moment of the bit point G touch with the π plane, it is instantaneous centre of the bit velocities, therefore, its absolute velocity equals zero. Then [25]

$$v_G^{abs} = v_C^{abs} + \Omega \times \rho = 0, \tag{10}$$

where v_C^{abs} is the absolute velocity of the C point of the bit, Ω is the vector of angular velocity of the reference frame $Cx_1y_1z_1$ fixed in the bit body.

Considering that the bit moves in the result of the DS rotation and elastic displacement and slewing of its lower end, one has

$$v_C^{abs} = (\dot{u}i + \dot{v}j) + \omega \times (ui + vj) = (\dot{u} - \omega v)i + (\dot{v} + \omega u)j,$$

$$\Omega = -\dot{v}'i + \dot{u}'j + \omega k,$$

$$\Omega \times \rho = \begin{vmatrix} i & j & k \\ -\dot{v}' & \dot{u}' & \omega \\ \rho_x & \rho_y & \rho_z \end{vmatrix} = (\dot{u}'\rho_z - \omega\rho_y)i + (\dot{v}'\rho_z + \omega\rho_x)j + (-\dot{v}'\rho_y - \dot{u}'\rho_x)k.$$

$$(11)$$

Substituting these correlations into Equation (10) and projecting onto the axes ox and oy, one can gain the kinematic boundary conditions for the bent DS at the C point

$$\dot{u} - \omega v + \dot{u}'\rho_z - \omega\rho_y = 0, \quad \dot{v} + \omega u + \dot{v}'\rho_z + \omega\rho_x = 0.$$

$$(12)$$

After substitution of the components of Equation (9) for ρ_x, ρ_y, ρ_z and some simplifications, the conclusive form of Equation (12) is constructed in the form:

$$\dot{u} - \omega v + \dot{u}'\sqrt{a^2 \sin^2\theta + b^2 \cos^2\theta} + \omega\frac{(b^2 - a^2)\sin\theta_x \cos\theta_x}{\sqrt{a^2 \sin^2\theta_x + b^2 \cos^2\theta_x}} = 0,$$

$$\dot{v} + \omega u + \dot{v}'\sqrt{a^2 \sin^2\theta + b^2 \cos^2\theta} + \omega\frac{(b^2 - a^2)\sin\theta_y \cos\theta_y}{\sqrt{a^2 \sin^2\theta_y + b^2 \cos^2\theta_y}} = 0.$$

$$(13)$$

The constraints described by these equations are nonholonomic as they are expressed through the derivatives of the unknown variables with respect to the t time [25] .

In the general case, the force boundary equations can be derived stemming from the theorem on the change of angular momentum of the bit. To exclude the unknown reaction of the contact interaction of the bit with the bore-hole bottom at the G point, it is chosen as the polar one. Then, these equations are formulated as follows:

$$\frac{\tilde{d}K_G}{dt} + \Omega \times K_G = M_G^{el}$$

$$(14)$$

where K_G is the bit moment momentum relative to the G point, M_G^{el} is the elastic forces moment with respect to the same point.

However, as shown in [24], the mass and inertia moment of the hollow bit are small in comparison with the inertia characteristics of the drill string and for this reason they can be neglected. If so, Equation (14) is simplified

$$M_G^{el} = 0. \tag{15}$$

The M_G^{el} moment is due to the vector of elastic moment

$$M_G^M = M_x i + M_y j + M_z k = EI\frac{\partial^2 v}{\partial z^2}i - EI\frac{\partial^2 u}{\partial z^2}j - M_z k \tag{16}$$

and vector of elastic forces

$$F = Q_x i + Q_y j - Tk = EI\frac{\partial^3 u}{\partial z^3}i + EI\frac{\partial^3 v}{\partial z^3}j - Tk. \tag{17}$$

The f vector generates

$$M_G^F = (-\rho)\times F = -\begin{vmatrix} i & j & k \\ \rho_x & \rho_y & \rho_z \\ Q_x & Q_y & -T \end{vmatrix} = -(-T\rho_y - Q_y\rho_z)i + (-T\rho_x - Q_x\rho_z)j - (Q_y\rho_x - Q_x\rho_y)k. \tag{18}$$

According to Equation (15), it can be written

$$M_G^M + M_G^F = 0. \tag{19}$$

After corresponding substitutions and projectings, one gains from (16), (18), (19)

$$EI\frac{\partial^2 v}{\partial z^2} + EI\frac{\partial^3 v}{\partial z^3}\rho_z + T\rho_y = 0,$$

$$-EI\frac{\partial^2 u}{\partial z^2} - EI\frac{\partial^3 u}{\partial z^3}\rho_z - T\rho_x = 0. \tag{20}$$

Taking into account Equation (9), transform Equation (20) to the form

$$\frac{\partial^2 u}{\partial z^2} + \frac{\partial^3 u}{\partial z^3}\sqrt{a^2\sin^2\theta + b^2\cos^2\theta} + \frac{T}{EI}\frac{(b^2 - a^2)\sin\theta_y\cos\theta_y}{\sqrt{a^2\sin^2\theta_y + b^2\cos^2\theta_y}} = 0,$$

$$\frac{\partial^2 v}{\partial z^2} + \frac{\partial^3 v}{\partial z^3}\sqrt{a^2\sin^2\theta + b^2\cos^2\theta} + \frac{T}{EI}\frac{(b^2 - a^2)\sin\theta_x\cos\theta_y}{\sqrt{a^2\sin^2\theta_x + b^2\cos^2\theta_x}} = 0. \tag{21}$$

Underline that the unknown cohesive force between the bit and bore-hole bottom is not included into these correlations. This effect is gained owing to the application of the nonholonomic approach to the problem and choice of the instantaneous centre of velocities (point G) as a polar one.

Boundary conditions (13), (21) represent the kinematic and quasi-static constraints imposed on the movement of the bit.

As a result, differential Equation (1) together with boundary conditions (3), (4), (13), and (21) make up the three-point boundary value problem for the bottom hole assembly including the bit. Its solution can be derived only by numerical methods.

STATEMENT OF THE CAUCHY PROBLEM FOR THE BOTTOM ASSEMBLY DYNAMICS

The above-formulated equations specify dynamic equilibrium of an ellipsoidal bit. They should be supplemented by initial conditions setting initial disturbance. Numerical solution of the stated problem is performed through the use of finite difference method with application of implicit scheme of integration in time t. To realize it, the span length l is divided into n finite difference segments $\Delta z = l/n$ and at every discrete time moment t_j and nodal point z_i, Equation (1) are replaced by their algebraic analogs

$$EI\frac{u_{i+2,j+1} - 4u_{i+1,j+1} + 6u_{i,j+1} - 4u_{i-1,j+1} + u_{i-2,j+1}}{\Delta z^4} - T\frac{u_{i+1,j+1} - 2u_{i,j+1} + u_{i-1,j+1}}{\Delta z^2}$$

$$-M_z\frac{v_{i+2,j+1} - 2v_{i+1,j+1} + 2v_{i-1,j+1} - v_{i-2,j+1}}{2\Delta z^3} - \gamma_t\,\omega^2 u_{i,j+1}$$

$$-2\gamma_t\,\omega\frac{v_{i,j+1} - v_{i,j-1}}{2\Delta t} + \gamma_t\frac{u_{i,j+1} - 2u_{i,j} + u_{i,j-1}}{\Delta t^2} = 0,$$

$$EI\frac{v_{i+2,j+1} - 4v_{i+1,j+1} + 6v_{i,j+1} - 4v_{i-1,j+1} + v_{i-2,j+1}}{\Delta z^4} - T\frac{v_{i+1,j+1} - 2v_{i,j+1} + v_{i-1,j+1}}{\Delta z^2}$$

$$+M_z\frac{u_{i+2,j+1} - 2u_{i+1,j+1} + 2u_{i-1,j+1} - u_{i-2,j+1}}{2\Delta z^3} - \gamma_t\,\omega^2 v_{i,j+1}$$

$$+2\gamma_t\,\omega\frac{u_{i,j+1} - u_{i,j-1}}{2\Delta t} + \gamma_t\frac{v_{i,j+1} - 2v_{i,j} + v_{i,j-1}}{\Delta t^2} = 0. \tag{22}$$

Here $u_{i-2,j+1} = u\big(z_i - 2\Delta z, t_j + \Delta t\big)$, $u_{i,j} = u\big(z_i, t_j\big)$, and so on.

The skeleton diagrams corresponding to the finite difference operators (22) are represented inFigure 5. It is considered that in time layers j and $j-1$, variables $u(i,j)$, $u(i,j-1)$, $v(i,j)$, $v(i,j-1)$ are known for $1 \le i \le n$ but variables $u(i,j+1)$, $v(i,j+1)$ $(1 \le i \le n)$ in layer $j+1$ are indeterminate. Then, Equation (22) together with the algebraized boundary conditions make it possible to formulate a system of linear algebraic equations in time layer $j+1$ for every time instant $t_{j+1} = t_j + \Delta t$. The right side of this system is expressed through the known varibles $u(i,j)$, $u(i,j-1)$, $v(i,j)$, $v(i,j-1)$ $(1 \le i \le n)$ calculated at previous time moments t_{j-1} and t_{j-2}. Solutions of these systems allow one to simulate the system motion by the step-by-step method. The accepted approach is effective for the considered problem owing to its numerical stability in respect to non-linear systems. But it is more cumbersome in connection with the necessity to solve a system of linear algebraic equations at every step of the time increment. At the same time, this approach is less accurate in comparison with the use of the explicit finite difference scheme. So, to achieve sufficient calculating precision, it is necessary to perform calculations with rather small values of the Δt increments.

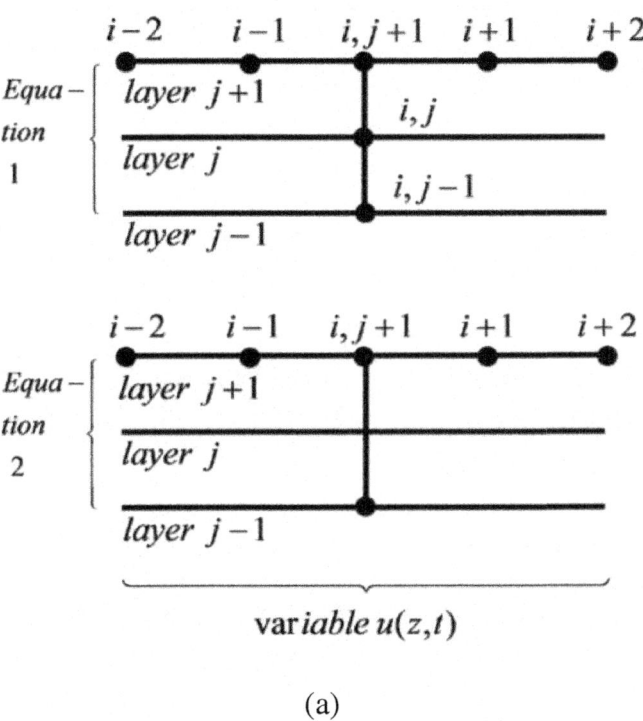

(a)

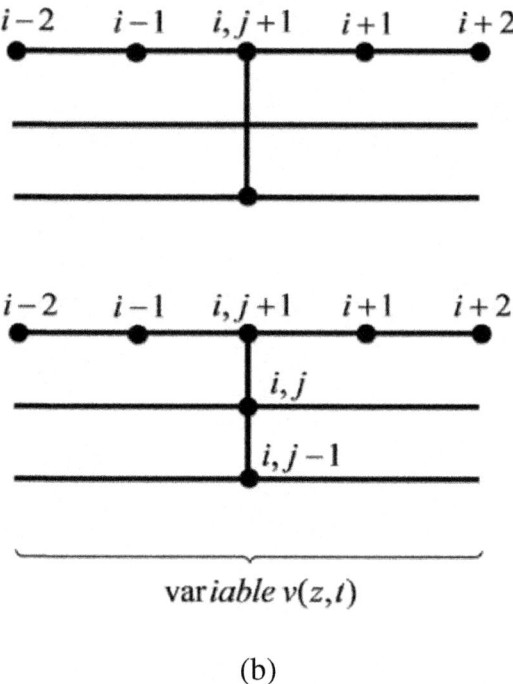

variable $v(z,t)$

(b)

Figure 5: Skeleton diagrams of finite difference operators of constitutive Equation (22).

THE SIMPLEST SCHEMES OF FORWARD AND BACK-WARD WHIRL ROLLINGS OF ROTATING ELLIPSOIDS ON AN UNEVEN PLANE

As is shown by experimental and theoretic investigations [1] - [4] , free rotating oblong ellipsoidal bodies with small geometric or mass distortions (Celtic stones) have the tendency to change the directions of their rotations. Analogous effects are typical to drill bits. They can also perform forward and backward whirlings and change their directions, describing the trajectories with the shapes of multipetal flowers cutting multilobed cross-sec- tions in bore-holes.(Thor Viggo et al. 1988), (Schen et al. 2005). But as distinguished from Celtic stones, they can have different geometry (including oblong and oblate ellipsoids, they are not free, they rotate with prescribed angular velocity ω , and change their orientation according to elastic bending of the DS axis and its tilting in vibrating.

Since the objective of this work is to understand the interaction between these geometric, kinematic and structural factors, the 3D model considering coupling of the DS bending with the ellipsoidal bit, rolling on the bore-hole bottom, is reckoned to be adequate. For a real DS structure, however, the results should only be interpreted in a qualitative sense.

To reveal the basic causes influencing on the modes of the bit motion on the bore-hole bottom, consider the simplest schemes of nonholonomic rolling of a rotary ellipsoid body on an uneven plane. Let the body be attached to an elastic rod rotating with angular velocity ω. For the sake of clearness, separate the motion states where the axes OX, Ox and OY, Oy are collinear and the plane cdg of the body inclination coincides with the xoz plane (Figure 6). Then, if the ellipsoid is elongated (b > a) and angle θ of its inclination to vertical Oz is positive, then at the considered time moment, the velocities of the point G of the body touch with the π plane (v_G^π) and of the ellipsoid apex D (v_D) are parallel to the Oy axis and the body moves in the direction of its rotation around the system axis (Figure 6(a)). So, this case corresponds to the forward whirl regime. However, the situation changes if the displacements $u(C)$, $v(C)$ are positive but the θ angle is negative (Figure 6(b)). In this event, the angular velocity of the bit rolling equals horizontal component of the angular velocity ω, the velocities v_D, v_G reverse their directions and the body begins to roll around the rod in the opposite direction, realizing backward whirl motion.

Kinematics of the oblate ellipsoid can be all the more complicated if the rotation axis of the deformed DS is located between the ellipsoid apex D and touch point g. Then, depending on the θ angle sign, the apex D can move in the direction of rotation while the touch point g does this in the opposite direction (Figure 6(c)), and, on the contrary, the D point can move in the rotation direction, whereas the g point does this in the opposite direction (Figure 6(d) for the backward whirl).

Thus, one can sec that depending on the shape of the bent DS, the whirling of the ellipsoidal bits can be both forward and backward, though for the oblate bits, the backward whirl is predominant.

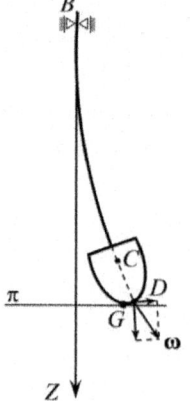

(a)

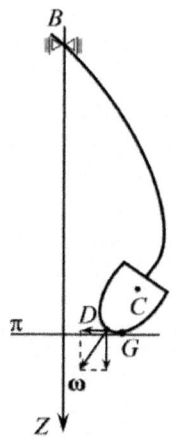

(b)

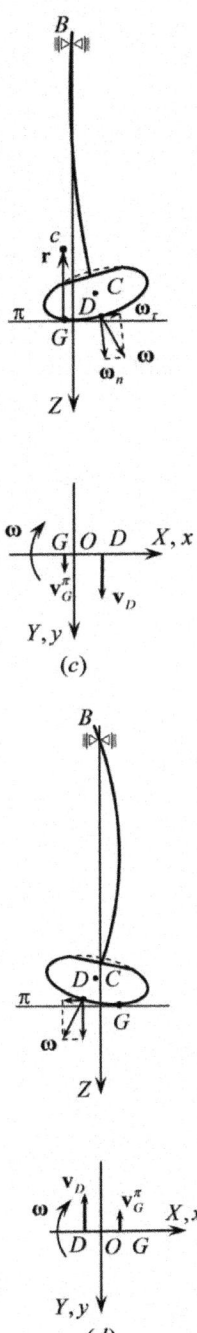

Figure 6: Kinematic schemes of forward and backward nonholonomic rollings of ellipsoidal bodies.

The elaborated model permits us to explain one further feature of the whirling process lying in the fact that it may acquire the most destructive modes when the angular velocity of the backward whirling essentially surpasses the rate ω of the DS and amounts from 5 to 30 times the speed of the DS [19] . For the purpose of corroborating this possibility refer to Figure 6(c). At the considered instant the bit is rolling with angular velocity ω_r about horizontal axis passing through its instantaneous center g of its velocities. Let point c be the curvature center of the bit surface section by the plane which contains point g and is normal to vector ω_r and r be the normal radius-vector constructed at point G. Then, its velocity v_c is normal to plane xoy and is equal to

$$v_c = \omega_r \times r = \omega_r \cdot r \cdot j.$$

As follows from [24] , the velocity v_G of the instantaneous center of velocities of the bit equals zero, but its trace in the π plane moves with the velocity

$$v_G^\pi = v_c.$$

Then

$$v_G^\pi = \omega_r \cdot r$$

and the angular velocity ω^{wh} of the bit whirling is

$$\omega^{wh} = \frac{v_c^\pi}{d} = \frac{\omega_r \cdot r}{d},$$

(23)

where d is the distance between the g point and axis OZ. It is evident that the ω^{wh} enlarges with the d decrease, but not to infinity as the ω_r reduces too. An important point also is that the whirl rate in Equation (23) depends on the r radius which is smaller in oblong bits and larger in oblate ones. Because of this, the oblate bits might be expected to be more predisposed to fast whirlings in comparison with oblong ones.

Kinematics of compound motion of the bit center c becomes more evident if to study it in the rotating coordinate system Oxyz (Figure 7).

In this case, the relative velocity vector v^r has the Cartesian components $\dot{x}i = \dot{u}i$ and $\dot{y}j = \dot{v}j$, so it can be expressed via the circumferential (v_{cir}^r) and radial (v_{rad}^r) components in the appropriate polar coordinate system.

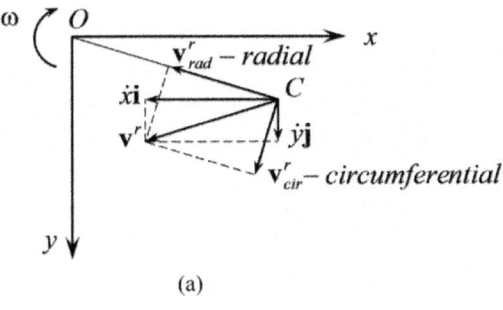

(a)

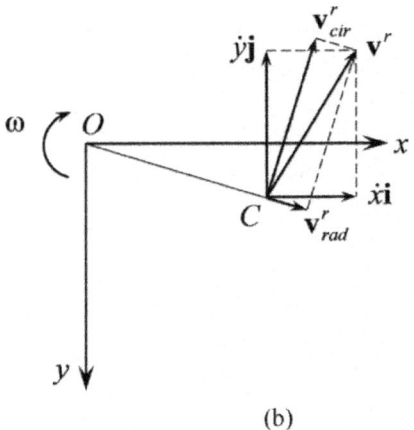

(b)

Figure 7: Top view of kinematic schemes of the relative velocity vector orientation.

Then the absolute velocity (v_C^{abs}) of the C point can be represented by the expression

$$v_C^{abs} = v^e + v^r,$$ (24)

where v^e is the bulk velocity vector calculated by the formula

$$v^e = \omega \times (xi + yj).$$

Thus, if vectors v^e and v_{cir}^r are oriented in the same direction, the bit rolls outstripping the DS rotation. If v^e and v_{cir}^r have different orientations, then the bit rolling lags behind the DS rotation when $|v^e| > |v_{cir}^r|$. It performs pure spinning without rolling when $|v^e| = |v_{cir}^r|$, and the bit rolling is opposite to the DS rotation when $|v^e| < |v_{cir}^r|$.

By virtue of the fact that in actual practice the bit possessing the ellipsoidal shape is subjected to the action of forces and moments from the elastic vibrating DS, it can permanently transfer from one kinematic scheme represented in Figure 6 and Figure 7 to another, changing directions of its rotational motions, as happens with the Celtic stones. In more complex cases, when these changes of directions occur repeatedly due to elastic vibration of the DS, the trajectories of the bit centre may represent more complicated figures closely resembling multipetal flowers.

SIMULATION OF WHIRL VIBRATIONS OF OBLONG AND OBLATE ELLIPSOIDAL BITS

The elaborated techniques were used for computer simulation of whirling vibrations of DSs with elongated and oblate ellipsoidal bits. In drilling practice, a great variety of determining factors have to be considered in technological designs. They differ essentially by the bore-hole diameters (up to 40 sm), drill string materials (steel, aluminum, titanium, composite), lengths of lower spans in bottom hole assembly (range from 9 to 18 m, shapes of bits, forces on bits (up to 10^6 N), torques on bits (up to 10^4 N·m), angular velocities of the DSs (more than 10 rad^{-1}), and others. In our investigations, the following typical values of the characteristic parameters were chosen for the analysis: E = 2.1 × 10^{11} Pa, ρ = 7.8 × 10^3 kg/m³, ρ_1 = 1.5 × 10^3 kg/m³, l = 9 m, e = 1 m, $F = \pi\left(r_1^2 - r_2^2\right) = 5.34 \times 10^{-3}$ m² , $I = \pi\left(r_1^4 - r_2^4\right)/4 = 1.94 \times 10^{-5}$ m⁴ , $F_l = \pi r_2^2 = 2.01 \times 10^{-2}$ m² , r_1 = 0.09 m, and r_2 = 0.08 m. Here, r_1 and r_2 are the outer and inner radii of the DS pipe; the values of a and b were varied.

As shown in [24] , the whirling regimes essentially depend on bending stiffness of the DS. With its reduction, the DS becomes flexible and the bit axis acquires additional opportunities to deviate from vertical state and to begin its whirling motions. If to take into account that the DS stiffness tends to zero with its bringing nearer to Eulerian critical state, then the expediency to evaluate proximity of the DS stress-strain state to critical one becomes evident. But the task of the DS stability analysis represents a complex independent multi parametric problem because the DS is prestressed by torque and variable axial force, it rotates, mud current moves inside it, and it performs longitudinal, lateral and torsional vibrations simultaneously. Inasmuch as this problem has not general solution, it is of interest to consider some simple partial combinations of loads, permitting to assess the DS stiffness. There are two sets of the rod loadings whose critical combinations can be determined analytically. For example, consider the case when a tubular rotating pinned rod is compressed by axial force t and a liquid is flowing with velocity v inside it. Then, critical values of parameters t, V, and ω are coupled by analytical correlation [32]

$$\frac{\pi^4}{l^2}EI + \pi^2 T_{cr} - l^2\gamma_t\omega_{cr}^2 - \pi^2\gamma_l V_{cr}^2 = 0.$$

(25)

Analytical solution of the Eulerian stability problem was received also for the case when the pinned rod was loaded by a torque M_z and axial force t:

$$T_{cr} = -\frac{\pi^2 EI}{l^2} + \frac{M_{z,cr}^2}{4EI}.$$

(26)

So then, Eqation (25) can be used for approximate evaluation of nearness of the lower segments of the DS to the critical states when torque value is low, Equation (26) is valid when the DS rotation and mud current can be neglected.

The techniques used for analysis of equation system (1), (3), (4), (12), and (21), as stated above, represent an implicit computational algorithm of integration. It is absolutely stable for any time increment Δt of the integration process but to provide the satisfactory accuracy, it should be fairly small. In practical computations, the value of Δt was determined experimentally by the trial-and-error method. In the present analysis the value $\Delta t = 10^{-4}$ s was used. To validate the computation accuracy, the calculations with the step $\Delta t = 10^{-5}$ s were performed. In these two cases, the integration results practically coincided.

To actuate starting motion of the system, it was assumed that small (1 cm in amplitude) elastic sinusoidal bending deformations were introduced into the shapes of the two lowest sections of the DS. Next, the DS (and bit) behavior was traced.

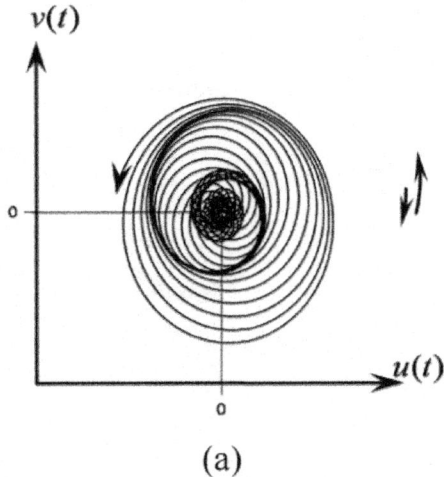

(a)

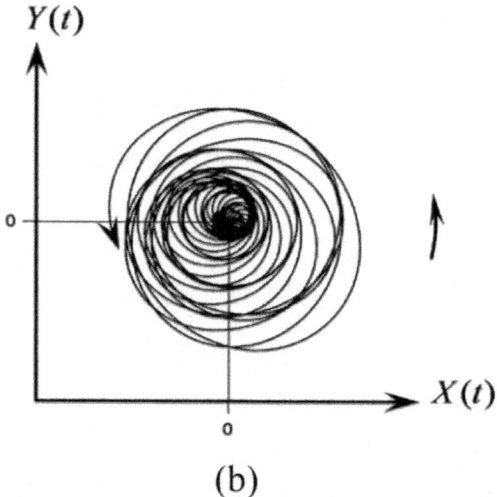

(b)

Figure 8: Bottom view of the motion trajectories of the oblong ellipsoidal bit center c in the rotating (a) and immovable (b) coordinate systems (a = 0.1 m, b = 0.3 m, ω = 5 rad/s, T = −5.6 × 10⁵ N, M_z = −1 × 10⁵ N·m).

The data of numerical analysis for the case of elongated ellipsoidal bit (a = 0.1 m, b = 0.3 m) are demonstrated in Figure 8 (ω = 5 rad/s) and Figure 9 (ω = 10 rad/s). In either example, the t forces are nearing the critical values (T_{cr} ≈ −7.8 × 10⁵ N for the first example and T_{cr} ≈ −7.3 × 10⁵ N for the second one).

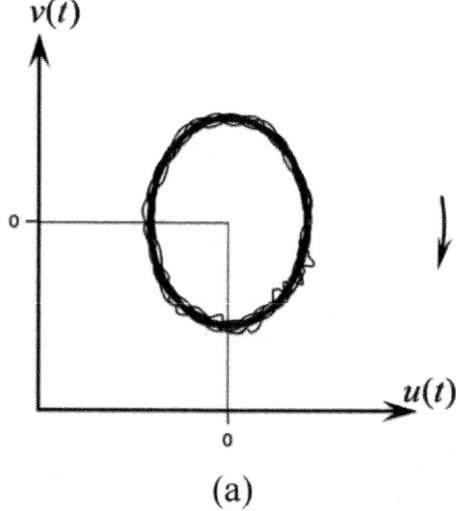

(a)

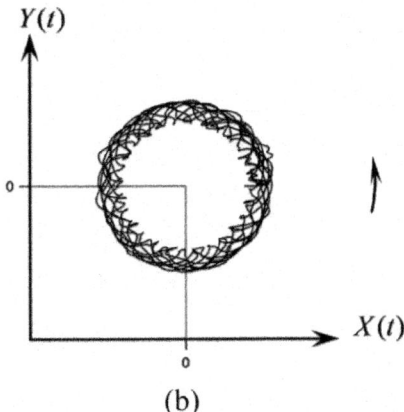

(b)

Figure 9: Bottom view of the motion trajectories of the oblong ellipsoidal bit center C in the rotating (a) and immovable (b) coordinate systems (a = 0.1 m, b = 0.3 m, ω = 10 rad/s, T = −6 × 10⁵ N, M_z = 0).

The trajectory in rotating coordinate system Oxyz shown in Figure 8(a) testifies that firstly the directions of the DS rotation and the bit motion were the same. But after some time, the bit began to describe very complicated curve ("to dance", as the Celtic stones did) and then changed orientation of its motion. Complicated trajectory was also formed in the immovable coordinate system (Figure 8(b)). Its parameters are determined with the use of formulae

$$X = u\cos\omega t - v\sin\omega t, \quad Y = u\sin\omega t + v\cos\omega t \tag{27}$$

It can be concluded that the considered regime is unstable as the motion curves have the shapes of expanding spirals.

However, enlargement of rotation velocity ω till 10 rad/s (Figure 9) stabilizes the system motion. After some disturbance, it traces elliptic paths in the rotating reference frame (Figure 9(a)) and circles with high frequency jitter distortions in the immovable one (Figure 9(b)). This regime, though stable, is also troublesome because it is connected with large accelerations and inertia forces.

Transition from elongated bits to oblate ones leads to complication of their motions. In Figure 10and Figure 11, the case a = 0.3 m, b = 0.1 m is represented for different values of ω, t and M_z. As shown in Figure 6, the oblate ellipsoid is characterized by larger lever between the central (c) and contact (g) points. Because of this, its whirling is accompanied by larger swings and more complicated modes of motions in the immovable reference frame (Figure 10(b) and Figure 11(b)). The motion mode shown in Figure 11(b) is of particular interest, since it exemplifies change of the whirl direction and its trajectory possesses a lot of loops (nearly cuspidal points) with enlarging accelerations and inertia forces.

Absolute motion of the oblate bit is more complicated even for small t values when its center c performs nearly stationary harmonic vibrations $u(t) = u_s \cos kt$, $v(t) = v_s \sin kt$ in the rotating coordinate system (Figure 12(a) for the case a = 0.3 m, b = 0.1 m, ω = 5 rad/s, T = -10^4 N, $M_z = -10^4$ N·m). The trajectory of its absolute motion in the immovable system OXYZ is represented in Figure 12(b). It can be seen that owing to change of the bit axis orientation (as shown in Figure 6(c) and Figure 6(d)), it periodically changes directions of its velocity, resulting in generating slow and fast motions with large accelerations and inertia forces. These forces can be dangerous for strength of the bit and its diamonds.

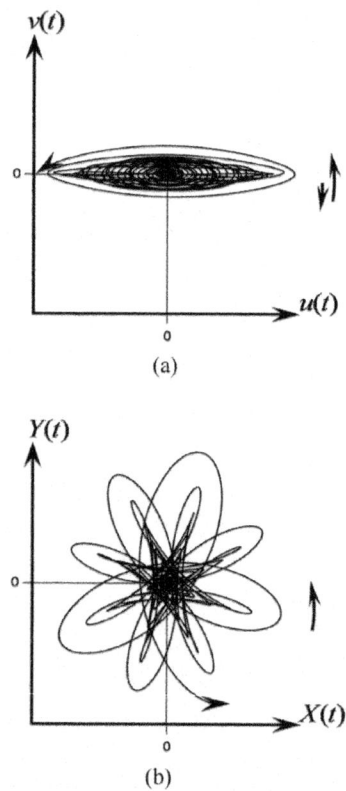

(a)

(b)

Figure 10: Bottom view of the motion trajectories of the oblate ellipsoidal bit center C in the rotating (a) and immovable (b) coordinate systems (a = 0.3 m, b = 0.1 m, ω = 10 rad/s, T = -5.8×10^5 N, $M_z = -1 \times 10^5$ N·m).

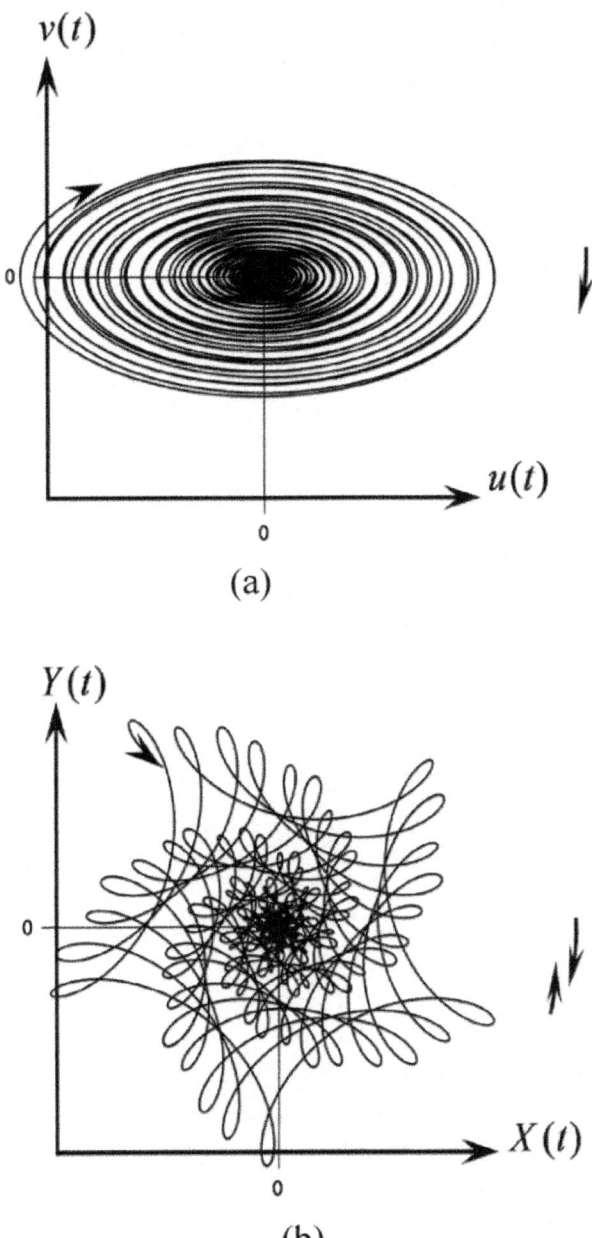

Figure 11: Bottom view of the motion trajectories of the oblate ellipsoidal bit center C in the rotating (a) and immovable (b) coordinate systems ($a = 0.3$ m, $b = 0.1$ m, $\omega = 20$ rad/s, $T = -5 \times 10^5$ N, $M_z = -5 \times 10^4$ N·m).

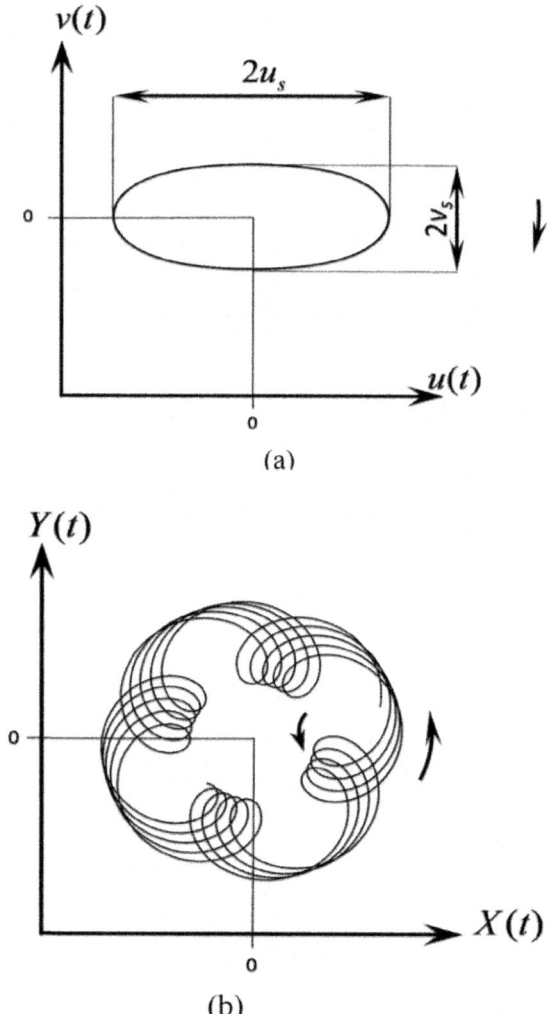

Figure 12: Bottom view of the motion trajectories of the oblate ellipsoidal bit center C in the rotating (a) and immovable (b) coordinate systems (a = 0.3 m, b = 0.1 m, ω = 5 rad/s, T = −1 × 10⁴ N, M_z = −1 × 10⁴ N·m).

The mode of the DS bending corresponding to the stage of comparatively slow motion is shown inFigure 13. It has a simple shape.

As noted above, the nonholonomic model advantage consists in its feature to study the bit motion without taking into consideration the cohesive force between contacting bodies. This force can be obtained from the condition of dynamic equilibrium after general solution of the stated problem. Really, the bit separated from the elastic DS moves in the horizontal plane under action of elastic force F^{el}, inertia force F^{in} and cohesive force F^{coh} which satisfy the equation

$$F^{el} + F^{in} + F^{coh} = 0. \tag{28}$$

If the problem is solved and forces F^{el} and F^{in} are calculated, the F^{coh} force can be found from the equality

$$F^{coh} = -F^{el} - F^{in}. \tag{29}$$

The F^{el} force stems from Equation (17)

$$F^{el} = EI \frac{\partial^3 u}{\partial z^3} i + EI \frac{\partial^3 v}{\partial z^3} j. \tag{30}$$

It follows from the above-mentioned notes that the mass of the hollow bit is small in comparison with the inertia characteristics of the DS and for this reason it can be neglected. Then, the resulting cohesive force can be expressed as follows:

$$F^{coh} = EI \sqrt{\left(\frac{\partial^3 u}{\partial z^3}\right)^2 + \left(\frac{\partial^3 v}{\partial z^3}\right)^2}. \tag{31}$$

Shown in Figure 14 is the diagram of the $F^{coh}(t)$ change constructed for the case presented inFigure 12. It can be seen that at the initial stage of the whirl process this function has a small jerk caused by initial perturbation but subsequently it transits to stationary regime in the limits $5500 < F^{coh} < 8700$ N. Taking into account that the axial force acting on the bit equals t = 10^4 N, one can infer (see Figure 2 and inequality (5)) that the regime of pure rolling without sliding can be realized if condition $\mu \geq F^{coh}/T = 0.87$ is fulfilled. Note that if the bit is new and its diamond spikes are sharp, the discussed conditions can be true. Because of this, in the considered case, the nonholonomic model use is justified.

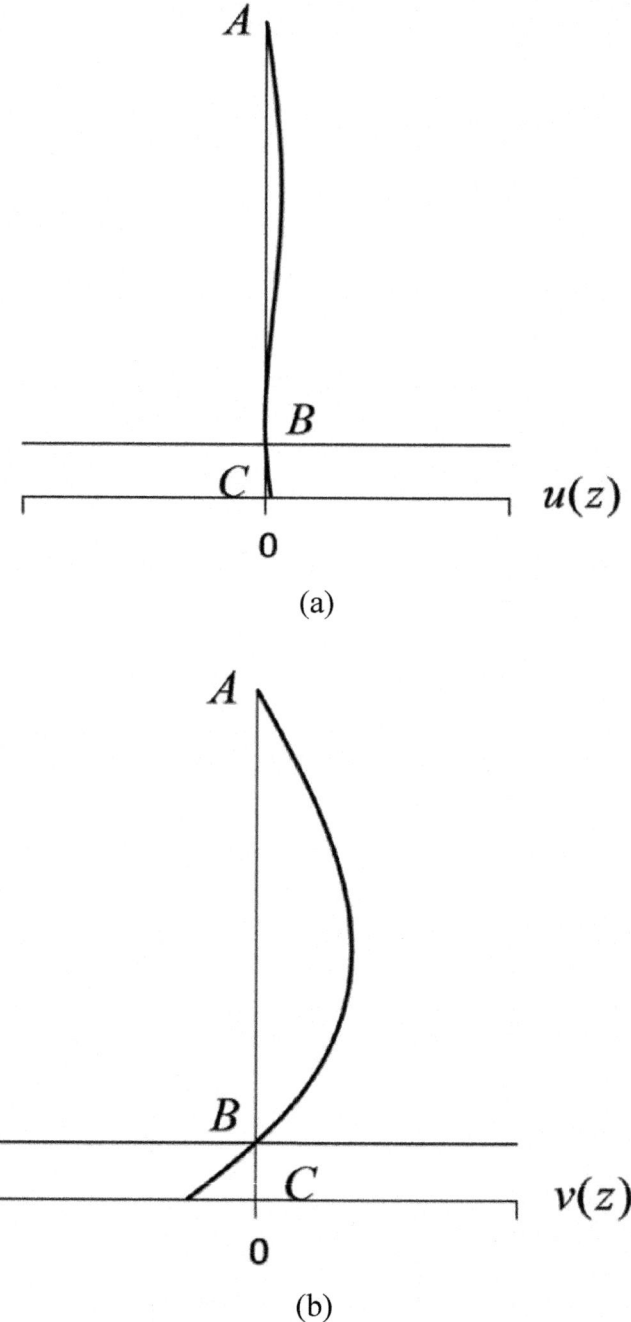

Figure 13: Modes of the DS bending in the rotating coordinate system (a = 0.3 m, b = 0.1 m, ω = 5 rad/s, T = −1 × 10^4 N, M$_z$ = −1 × 10^4 N·m).

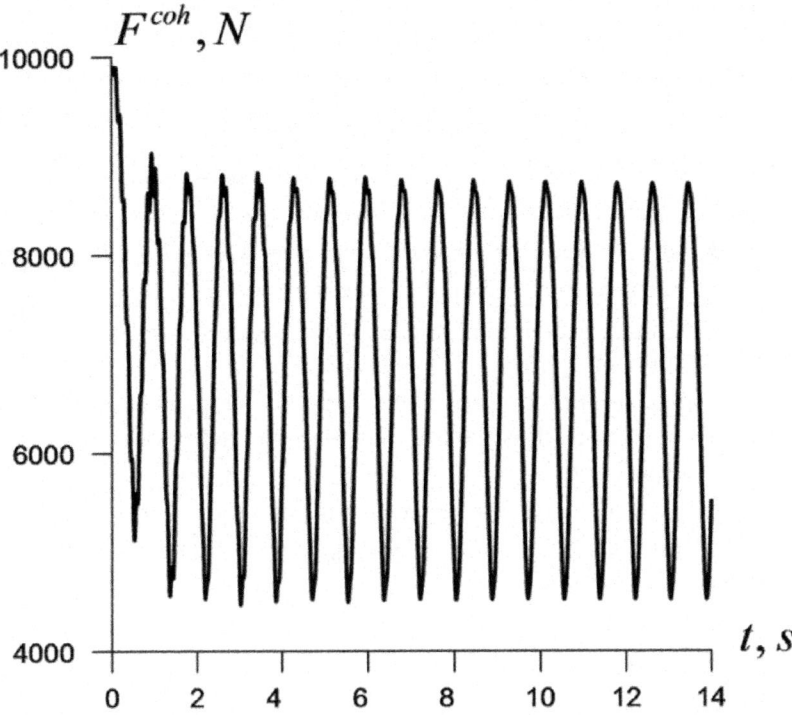

Figure 14: The diagram of the cohesive force change (a = 0.3 m, b = 0.1 m, ω = 5 rad/s).

CONCLUSIONS

1. In the paper, the problem on rolling of a rigid ellipsoidal drill bit on a bore-hole bottom plane is stated. The bit is supposed to be attached to the lower end of an elastic rotating drill string. Two mechanical models of the bit motions based on assumption of the possibilities of its pure rolling whirling and rolling with sliding are discussed. The nonholonomic dynamic model is taken for analysis. An analogy between the phenomena of nonholonomic motions of the Celtic stones and the ellipsoidal bits is discussed. The techniques for solution of the equations of the bit whirling are proposed.

2. With the use of the elaborated approach, the computer simulation of whirling vibrations of the initially disturbed systems is performed. The oblong and oblate ellipsoidal shapes of the bits are chosen for analysis. As demonstrated by theoretic analysis of the simplest kinematic schemes of the ellipsoid bit rolling, their whirlings can be both forward and

backward, though for the oblate bits, the backward whirl is predominant. As established by calculations under conditions of reduced bending stiffness of the drill string, the bit whirling can be stable as well as unstable. With the use of the simplest kinematic schemes of the ellipsoid body rolling, it is demonstrated that the oblate bits are more sensitive to initial disturbances and trajectories of their whirling paths can have loops and cuspidal points corresponding to bendings of the DSs and their bit axes tiltings these regimes are connected with enlargement of acceleration values and appropriate inertia forces.

3. The distinctive characteristic of the considered regimes is the found possibility to stop the circumferential motion of the bit whirling and to change its direction to inverse one, as happens with nonholonomic whirling of free ellipsoidal bodies with geometric or inertial distortions (the Celtic stones).

4. The elaborated techniques can be used for prediction and qualitative evaluation of the ellipsoid bit whirling regimes at the stages of the borehole design and drivage.

REFERENCES

1. Walker, G.T. (1895) On a Curious Dynamical Property of Celts. Proceedings of the Cambridge Philosophical Society, 8, 305-306.

2. Walker, J. (1979) The Mysterious "Rattleback": A Stone That Spins in One Direction and Then Reverses. Scientific American, 241, 144-149.

3. Lindberg, R.E. and Longman, R.W. (1983) On the Dynamic Behavior of the Wobblestone. Acta Mechanica, 49, 81-94. http://dx.doi.org/10.1007/BF01181756

4. Pascal, M. (1983) Asymptotic Solution of the Equations of Motion for a Celtic Stone. Journal of Applied Mathematics and Mechanics, 47, 269-276. http://dx.doi.org/10.1016/0021-8928(83)90016-3

5. Jansen, J.D. (1992) Whirl and Chaotic Motion of Stabilized Drill Collars. SPE Drilling Engineering, 7, 107-114. http://dx.doi.org/10.2118/20930-PA

6. Christoforou, A.P. and Yigit, A.S. (1997) Dynamic Modeling of Rotating Drillstrings with Borehole Interactions. Journal of Sound and Vibration, 206, 243-260. http://dx.doi.org/10.1006/jsvi.1997.1091

7. Leine, R.I., Van Campen, D.H. and Keulties, W.J.G. (2002) Stick-Slip Whirl Interaction in Drillstring Dynamics. Journal of Vibration and Acoustics, 124, 209-220. http://dx.doi.org/10.1115/1.1452745

8. Gulyaev, V.I., Khudolii, S.N. and Borshch, E.I. (2010) Whirl Vibrations of the Drillstring Bottom Hole Assembly. Strength of Materials, 42, 637-646. http://dx.doi.org/10.1007/s11223-010-9252-y

9. Brett, J.F., Warren, T.M. and Behr, S.M. (1990) Bit Whirl—A New Theory of PDC Bit Failure. SPE Drilling Engineering, 5, 275-281. http://dx.doi.org/10.2118/19571-PA

10. Langeveld, C.J. (1992) PDC Bit Dynamics. SPE/IADC Drilling Conference, New Orleans, 18-21 February 1992, 227-242. http://dx.doi.org/10.2118/23867-ms

11. Schen, A.E., Snell, A.D. and Stanes, B.H. (2005) Optimization of Bit Drilling Performance Using a New Small Vibration Logging Tool. SPE/IADC Drilling Conference, Amsterdam, 23-25 February 2005. http://dx.doi.org/10.2118/92336-ms

12. Aarrestad, T.V. and Kyllingstad, A. (1988) An Experimental and Theoretical Study of a Coupling Mechanism between Longitudinal and Torsional Drillstring Vibrations at the Bit. SPE Drilling Engineering, 3, 12-18. http://dx.doi.org/10.2118/15563-PA

13. Chen, S.L., Blackwood, K. and Lamine, E. (2002) Field Investigation of the Effects of Stick-Slip, Lateral, and Whirl Vibrations on Roller-Cone Bit Performance. SPE Drilling & Completion, 17, 15-20. http://dx.doi.org/10.2118/76811-PA

14. Wu, X., Paez, L.C., Partin, U.T. and Agnihotri, M. (2010) Decoupling Stick/Slip and Whirl to Achieve Breakthrough in Drilling Performance. IADC/SPE Drilling Conference and Exhibition, New Orleans, 2-4 February 2010. http://dx.doi.org/10.2118/128767-ms

15. Ledgerwood, L.W., Jain, J.R., Hoffmann, O.J. and Spencer, R.W. (2013) Downhole Measurement and Monitoring Lead to an Enhanced Understanding of Drilling Vibrations and Polycrystalline Diamond Compact Bit Damage. SPE Drilling & Completion, 28, 254-262. http://dx.doi.org/10.2118/134488-PA

16. Ledgerwood, L.W., Hoffmann, O.J., Jain, J.R., El Hakam, C., Herbig, C., Spencer, R.W. and Hughes, B. (2010) Downhole Vibration Measurement, Monitoring, and Modeling Reveal Stick-Slip as a Primary Cause of PDC-Bit Damage in Today's Applications. SPE Annual Technical Conference and Exhibition, Florence, 19-22 September 2010, 2652-2661. http://dx.doi.org/10.2118/134488-MS

17. Sowers, S.F., Dupriest, F.E., Bailey, J.R. and Wang, L. (2009) Roller Reamers Improve Drilling Performance in Wells Limited by Bit and

Bottomhole Assembly Vibrations. SPE/IADC Drilling Conference and Exhibition, Amsterdam, 17-19 March 2009. http://dx.doi.org/10.2118/119375-ms

18. Johnson, S.Ch. (2008) A New Method of Producing Laterally Stable PDC Drill Bits. SPE Drilling & Completion, 23, 314-324. http://dx.doi.org/10.2118/98986-PA

19. Stroud, D., Pagett, J. and Minett-Smith, D. (2011) Real-Time Whirl Detector Improves RSS Reliability, Drilling Efficiency. Hart Exploration & Production Magazine, 84, 42-43.

20. Kovalyshen, Y. (2012) A New Model of Bit Whirl. IADC/SPE Asia Pacific Drilling Technology Conference and Exhibition, Tianjin, 9-11 July 2012. http://dx.doi.org/10.2118/156240-ms

21. Spanos, P.D., Chevallier, A.M. and Politis, N.P. (2002) Nonlinear Stochastic Drill-String Vibrations. Journal of Vibration and Acoustics, 124, 512-518. http://dx.doi.org/10.1115/1.1502669

22. Ritto, T.G., Soize, C. and Sampaio, R. (2009) Non-Linear Dynamics of a Drill-String with Uncertain Model of the Bit-Rock Interaction. International Journal of Non-Linear Mechanics, 44, 865-876. http://dx.doi.org/10.1016/j.ijnonlinmec.2009.06.003

23. Samuel, R. (2010) Friction Factors: What Are They for Torque, Drag, Vibration, Bottom Hole Assembly and Transient Surge/Swab Analyses? Journal of Petroleum Science and Engineering, 73, 258-266. http://dx.doi.org/10.1016/j.petrol.2010.07.007

24. Gulyayev, V.I. and Shevchuk, L.V. (2013) Nonholonomic Dynamics of Drill String Bit Whirling in a Deep Bore-Hole. Journal of Multi-Body Dynamics, 227, 234-244. http://dx.doi.org/10.1177/1464419313482658

25. Neimark, Ju.I. and Fufaev, N.A. (1972) Dynamics of Nonholonomic Systems. Translation of Mathematical Monographs, 33, 519 p.

26. Gulyayev, V.I., Glushakova, O.V. and Hudoliy, S.N. (2010) Quantized Attractors in Wave Models of Torsion Vibrations of Deep-Hole Drill Strings. Mechanics of Solids, 45, 264-274. http://dx.doi.org/10.3103/S0025654410020123

27. Gulyayev, V.I. and Glushakova, O.V. (2011) Large-Scale and Small-Scale Self-Excited Torsional Vibrations of Homogeneous and Sectional Drill Strings. Interaction and Multiscale Mechanics, 4, 291-311. http://dx.doi.org/10.12989/imm.2011.4.4.291

28. Gulyayev, V.I., Hudoliy, S.N. and Glushakova, O.V. (2011) Simulation of Torsion Relaxation Auto-Oscillations of Drill String Bit with Viscous and

Coulombic Friction Moment Models. Journal of Multi-Body Dynamics, 225, 139-152. http://dx.doi.org/10.1177/1464419311405571

29. Gulyayev, V.I. and Tolbatov, E.Yu. (2002) Forced and Self-Excited Vibrations of Pipes Containing Mobile Boiling Fluid Clots. Journal of Sound and Vibration, 257, 425-437. http://dx.doi.org/10.1006/jsvi.2002.5045

30. Gulyayev, V.I. and Tolbatov, E.Yu. (2004) Dynamics of Spiral Tubes Containing Internal Moving Masses of Boiling Liquid. Journal of Sound and Vibration, 274, 233-248. http://dx.doi.org/10.1016/j.jsv.2003.05.013

31. Gulyayev, V.I. and Borshch, O.I. (2011) Free Vibrations of Drill Strings in Hyper Deep Vertical Bore-Wells. Journal of Petroleum Science and Engineering, 78, 759-764. http://dx.doi.org/10.1016/j.petrol.2011.09.001

32. Gulyayev, V.I., Gaidaichuk, V.V., Solovjov, I.L. and Gorbunovich, I.V. (2009) The Buckling of Elongated Rotating Drill Strings. Journal of Petroleum Science and Engineering, 67, 140-148. http://dx.doi.org/10.1016/j.petrol.2009.05.011

Chapter 6

DRY FRICTION WITH VARIOUS FRICTIONS LAWS: FROM WAVE MODULATED ORBIT TO STICK-SLIP MODULATED

Paul Ndy Von Kluge, Djuidjé Kenmoé Germaine, and Kofané Timoléon Crépin

Department of Physics, Mechanics Laboratory, Faculty of Sciences, University of Yaoundé 1, Yaoundé, Cameroon

ABSTRACT

Choices of excitation signals are important in engineering sciences and in physical simulations; a sufficient excitation can be critical in modelling a complicated nonlinear dynamic system. The discontinuous dynamic of a non-linear, friction-induced with two idealized periodical forced oscillators is studied. The dry friction in the system follows the classical Coulomb law, and various friction characteristics of dry friction laws in engineering sciences. To capture the presence of the two driving forces, the system must be studied as a function of their frequency-modulated and its equivalent amplitude modulated waveforms. Our numerical investigation shows a rich dynamical behaviour including periodic, quasi-periodic motions, thus a variable dynamics phenomenon among others; such as modulated waves, modulated stick-slip, periodic oscillation, and periodic stick-slip. It seems that such excitation forces can be used to conveniently identify the existence of nonlinearity, dry friction effects, and strength degradation in the system. The results achieved via the Coulomb's law are compared with those obtained via two others particular friction laws: the complete model with Stribeck effect and Coulomb viscosity.

INTRODUCTION

In general, there are many different types of dry friction models and it is crucial to appropriately choose one which best suits to the modelled problem, however choices of excitation forces have been widely studied for the modal testing community [1] . In system identification, the need to investigate proper excitation forces has been recognized even for a system without memory [2] . For many years the topic of dry friction has been actively researched with many attempts to identify the causes of unwanted behaviour [3] - [5] . Friction-induced vibration, chatter and squeal [6] cause serious problems in many industrial applications, including turbine blade joints, robot joints, electric motor drives, water-lubricated bearings in ships and submarines, wheel/rail coupling of mass transit systems, machine tool/work piece systems and brake systems. These forms of vibration [7] can cause excessive wear of components: surface damage, fatigue failure, and noise.

From the mathematical point of view, the appearance of the discontinuous differential equations is usual, where the character of this discontinuity depends on the friction character adopted [8] , but seeking an effective means of decoupling the complexity of excitations from that of the nature of the system is important. The presence of Coulomb friction makes the analysis of the problem challenging due to the dependence of the friction force on the velocity in a complicated and discontinuous manner. If one considers dynamics of the system where the relative velocity practically remains constant, there is no need for sophisticated dry friction models and even the easiest one described by the Coulomb law will be enough. However in many cases, the variation of relative velocity is large and often the velocity changes its sign [9] . But when adding other particular friction laws such as complet model with Stribeck effect or Coulomb + viscosity, other complicated phenomenons occur. Systems with dry friction possess many different types of dynamical behaviour [10] , such as periodical oscillation and periodical stick-slip. An approximate analysis of stick-slip vibration is based on dividing the motion into two phases: a stick phase in which the velocity of the motion is constant and a slip phase in which the motion is approximated as a circular (pure sinusoidal) motion with constant amplitude. The solutions of the two phases are batched together. Applying oscillation force sometimes to control the system moves the system from stick-slip to smooth sliding.

The "mass on belt system" falls within the category of hybrid system or switching system [11] [12] . The switching behaviour is brought in by the play of dry friction originating between the surface of the mass and the belt. The mass can either be in a stick mode or a slip mode and this creates a switching type system. Friction induced and self-excitation oscillations are affected by high-frequency external excitations [13] . Friction properties change under the action of fast vibration. Generally, high-frequency excitations can reduce or totally suppress friction-induced vibrations [14] . This study considers how such friction properties in two contact surfaces are changed in the presence of two external excitations. Among the vibration characteristics used, the vibration amplitude is especially important because it can directly affect stresses and thus the life of the system. The friction model is a sign function plus linear and cubic function of the relative velocity [15] . In the present work, we further our study into the dynamics of a single-degree-of-freedom system with dry friction and two harmonic driving forces. The aim is to understand how the response of the system is affected by the amplitude modulated and the ratio of the frequencies modulated of the two driving forces using either the Coulomb's law or other particular friction laws.

MODEL DESCRIPTION AND EQUATION OF THE MOTION

Model Description

The system under investigation is shown in Figure 1, where a mass m is connected to a fixed support via a linear spring with stiffness k and a viscous damper coefficient c. This oscillator slides or rests on the horizontal belt surface travelling with a constant speed v_0. Consider two harmonic driving forces exerting on the mass with the same amplitude p_0, but different frequencies w_1 and w_2, these forces are defined as

$$F_e(t) = p_0 \cos w_1 t + p_0 \cos w_2 t \tag{1}$$

The two excitation frequencies are proportional to each other in such a way that $w_2 = v w_1$ (if $v = 1$, the force degenerates to a single excitation case). The dry friction between the interfaces can follow Coulomb's law (characterized by the static coefficient μ_s and the smaller kinetic coefficient μ_k) or other frictions laws. Since the mass contacts the moving belt with friction, the mass can move along or rest on the belt. The equation of motion for such a friction-induced oscillator is:

$$m\frac{d^2x}{dt^2} + c\frac{dx}{dt} + kx = -F + F_e(t) \tag{2}$$

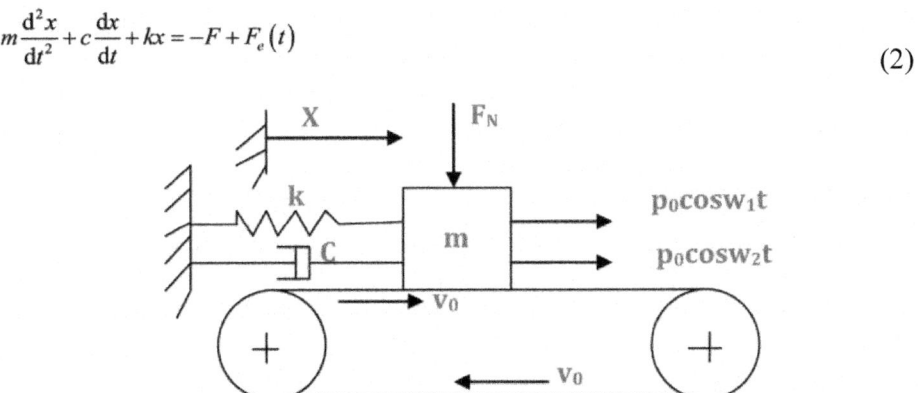

Figure 1: Model of the single-degree-of freedom friction oscillator excited by moving base and two driving forces.

where x denotes the displacement of the mass. F_N is the normal force in the contact area, i.e. The weight of the block $F_N = mg$.

With a non smooth transition, the resulting motion also shows a non smooth behaviour. The stick-slip systems belong to the class of non smooth systems, such as systems with stops, impacts, backlash or hysteresis. The nature of dynamic friction forces developed between bodies in contact is extremely complex and affected by a long list of factors: the constitution of the interface, the time scales and the frequency of the contact, the response of the interface to normal forces, inertia and thermal effects [16] , roughness of the contacting surfaces [17] . The dry friction force F depends on the relative speed of the surfaces in contact and can be defined as:

$$F = \mu F_N \, \text{sgn}\left(\frac{dx}{dt} - v_0\right) - a_1\left(\frac{dx}{dt} - v_0\right) + a_3\left(\frac{dx}{dt} - v_0\right)^3$$

(3)

where a_1 and a_3 are friction coefficient, and

$$\mu = \begin{cases} \mu_k, & \dfrac{dx}{dt} \neq v_0 \\[2mm] \mu_s, & \dfrac{dx}{dt} = v_0 \end{cases}$$

(4)

The relative velocity of the contact can be defined as: $v_r = \dfrac{dx}{dt} - v_0 = \dot{x} - v_v$ and

$$\text{sgn}(v_r) \begin{cases} = 1 & \text{if } (v_r > 0) \\ \in [-1,1] & \text{if } (v_r = 0) \\ = -1 & \text{if } (v_r > 0) \end{cases}$$

(5)

The significant role of various dry friction laws in engineering sciences can be illustrated in Figure 2. The friction characteristic described with Coulomb law is shown in Figure 2(a). The friction model does not specify the friction force for zero velocity. Figure 2(b) shows the case of coulomb + viscosity while Figure 2(c) illustrates the friction with Stribeck effect.

Equation of Motion

The equation of motion Equation (2) can be normalized using:

$$u_0 = \frac{p}{mw_0^2}, \quad x_f = \frac{\mu F_N}{k} = \begin{cases} x_{fk}, \dot{x} \neq v_0 \\ x_{fs}, \dot{x} = v_0 \end{cases}, \quad \eta = \frac{w_1}{w_0}, \quad w_0 = \sqrt{\frac{k}{m}}, \quad \tau = w_0 t,$$

$$\lambda = \frac{c}{mw_0}, \quad V_v = \frac{v_0}{w_0}, \quad b_1 = \frac{a_1}{mw_0^2}, \quad b_3 = \frac{a_3}{mw_0^2}, \quad \frac{dx}{d\tau} = \dot{x}, \quad \frac{d^2x}{d\tau^2} = \ddot{x}$$

(6)

where τ is dimensionless time coordinate, w_0 is the frequency of free oscillations of the mass.

Equation (2) can be rewritten then in a dimensionless form as:

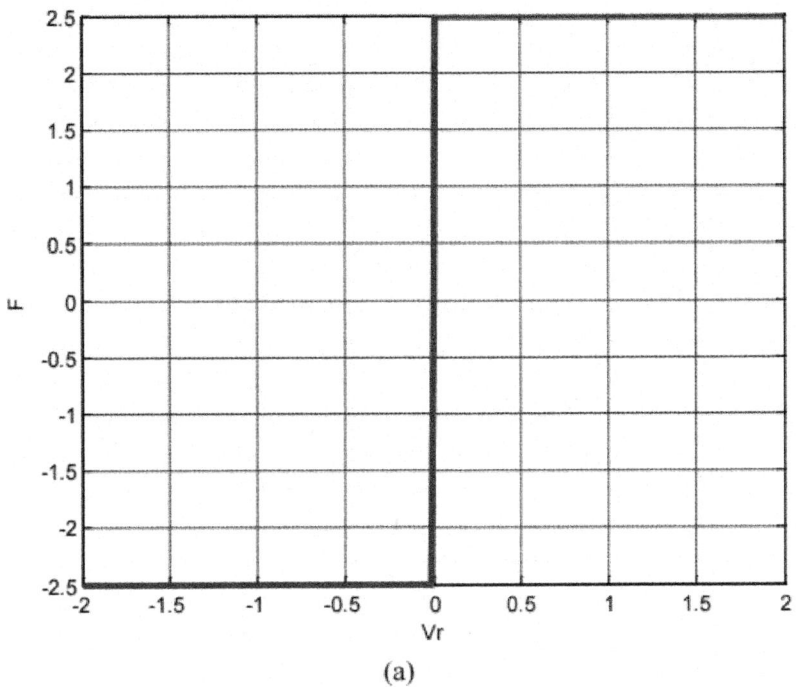

(a)

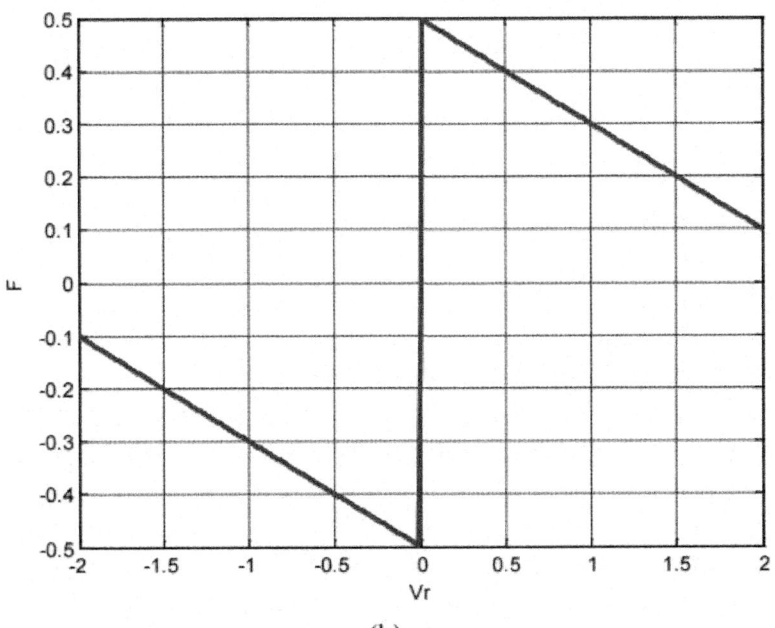

(b)

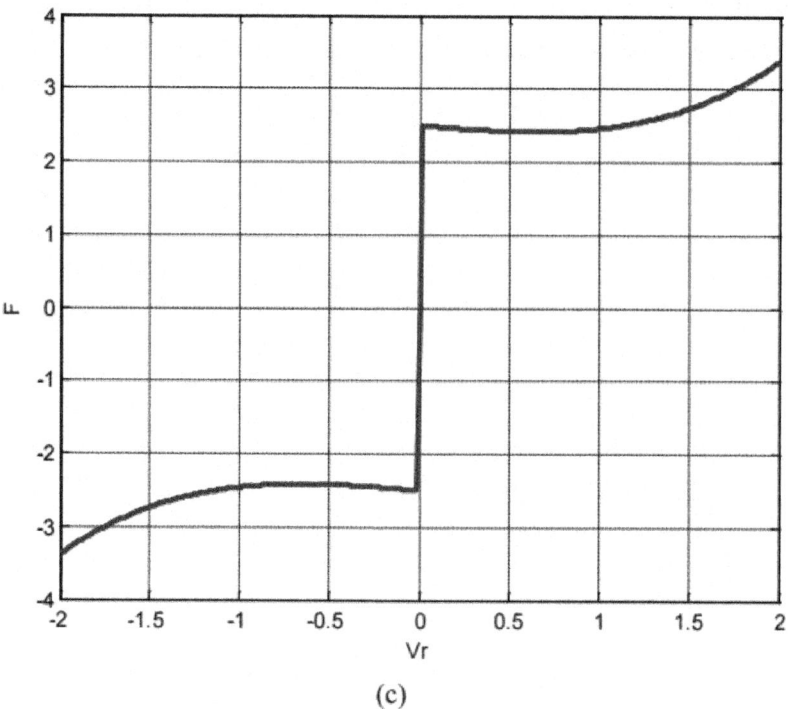

(c)

Figure 2: Velocity-dependent frictions laws giving by Equation (3): (a) non-smooth Coulomb law: ($a_1 = 0.0$, $a_3 = 0.0$); (b) Coulomb + viscosity: ($a_1 = 0.2$, $a_3 = 0.0$); (c) Complete model with Stribeck effect: ($a_1 = 0.2$, $a_3 = 0.16$).

$$\ddot{x} + \lambda\dot{x} + x = -x_f \, \mathrm{sgn}(\dot{x}-V_v) + b_1(\dot{x}-V_v) - b_3(\dot{x}-V_v)^3 + u_0 \cos(\eta\tau) + u_0 \cos(\nu\eta\tau) \quad (7)$$

The choice of one of the friction laws shown in Figure 2 can be obtained by monitoring the parameters b_1 and b_3 (dimensionless friction coefficient).

DYNAMIC RESPONSES

Analytical Treatment of the Motion of the Block

In the case where $b_1 = b_3 = 0$, one can treat Equation (7) analytically to show amplitude motion which occurs in the sliding phase. Assume that the motion starts when the block sticks on the belt, $\dot{x} = V_v$. The general form of the solution of Equation (7) can be write as $x = Ke^{-\frac{\lambda\tau}{2}}(c_1 \cos p\tau + c_2 \sin p\tau)$. It is the sum of the homogenous solution and the particular solution. If $x = A\cos\eta\tau + B\cos\nu\eta\tau$ is the particular solution, the final solution becomes.

$$x(\tau) = \mp x_{fk} + e^{-\frac{\lambda(\tau-\tau_0)}{2}}\left[c_1 \cos p(\tau-\tau_0) + c_2 \sin p(\tau-\tau_0)\right] + A\cos(\eta\tau - \theta_1) + B\cos(\nu\eta\tau - \theta_2) \quad (8)$$

where

$$c_1 = x_0 \pm x_{fk} - A\cos(\eta\tau_0 - \theta_1) - B\cos(\nu\eta\tau_0 - \theta_2)$$

$$c_2 = \frac{1}{p}\left[V_0 + \frac{\lambda}{2}c_1 + A\eta\sin(\eta\tau_0 - \theta_1) + B\nu\eta\sin(\nu\eta\tau_0 - \theta_2)\right] \quad (9)$$

With the initial conditions

$$x(\tau_0) = x_0, \quad \frac{dx}{d\tau} = V_0, \quad p = \sqrt{1 - \frac{\lambda^2}{4}}, \quad \theta_1 = \tan^{-1}\left(\frac{\lambda\eta}{1-\eta^2}\right), \quad \theta_2 = \tan^{-1}\left(\frac{\lambda\nu\eta}{1-\nu^2\eta^2}\right),$$

$$A = \frac{u_0}{\sqrt{(1-\eta^2)^2 + \lambda^2\eta^2}}, \quad B = \frac{u_0}{\sqrt{(1-\nu^2)^2 + \nu^2\lambda^2\eta^2}}. \quad (10)$$

It is noted that the upper and lower part of the compound signs in Equation (8) corresponds to the case of $\dot{x} > V_\nu$ and $\dot{x} < V_\nu$, respectively, c_1 and c_2 are constants.

An important idea associated with the response of an oscillator to the periodic force is resonance i.e., when the natural frequency of the oscillator is equal to the frequency of the periodic force. Figure 3 displays the evolution of the amplitude as a function of the parameter η. The curves of amplitude versus the ratio η show a resonance near the value $\eta = 1$, i.e., when the frequency w_1 is near the natural frequency w_0. The height of the peak depends on the applied u_0, the damped coefficient λ, as well as on the ratio of the force excitation frequencies ν. For smaller values of λ in Figure 3(c), the current frequency curve is sharply peaked, but for large values of λ the curve is flat.

Numerical Treatment

We further our study, in order to know how the response of the system is affected by the two harmonic driving forces. Numerical investigation will be done firstly, by using Coulomb's law and secondly with numerical investigation by assuming two others particular friction laws as mention above.

Numerical Analysis of the Motion of the Block

In order to understand dynamical processes in our model, Equation (7) has been integrated numerically using fourth-order Runge-Kutta scheme with time step $\Delta t = 0.04$. The results of this numerical investigation are shown in Figures 4-7, in which we obtained unless otherwise specified [18], for $\lambda = 1/2$, $\eta = 3/5$, $V_\nu = 1$, $x_{fk} = 2.5$, $x_{fs} = 4.0$, $b_1 = b_3 = 0$.

Through a series of numerical simulations, it seems that such excitation forces can be used to conveniently reproduce the stick-slip behaviour of ice streams in glaciology [19] . However, in the system, we observe two types of modulation: simple wave modulation and stick-slip modulation. Acronyms like "WM" and "SSM" stand for "wave modulated" and "stick-slip modulated", respectively. Depending on the amplitude u_0 and the ratio of the two frequencies ν of the external forces, the dynamical behaviour of the block can be classified in one of the following three categories:

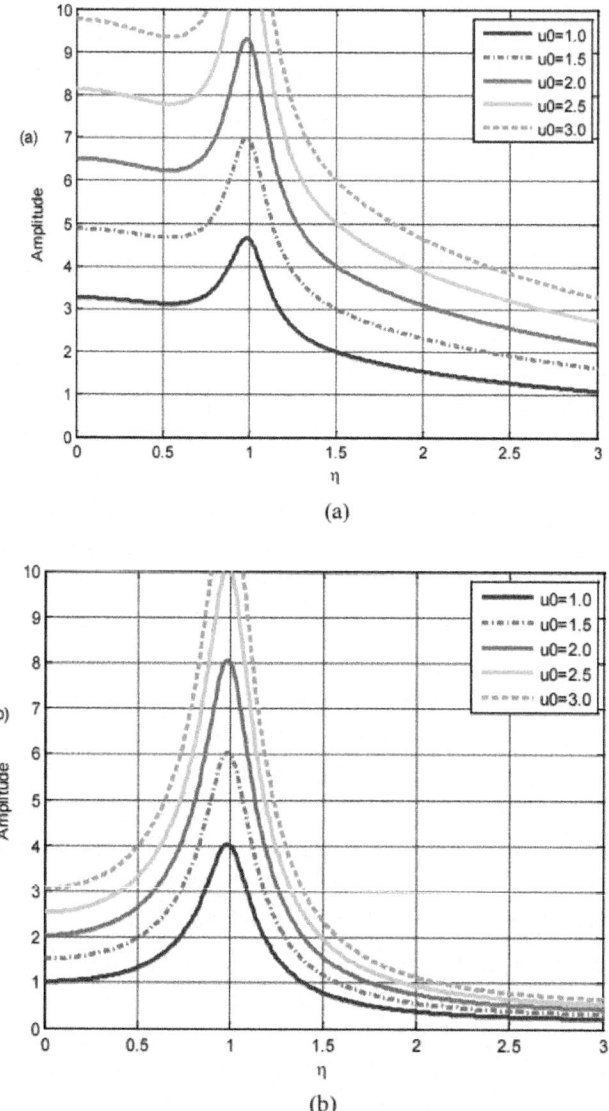

(a)

(b)

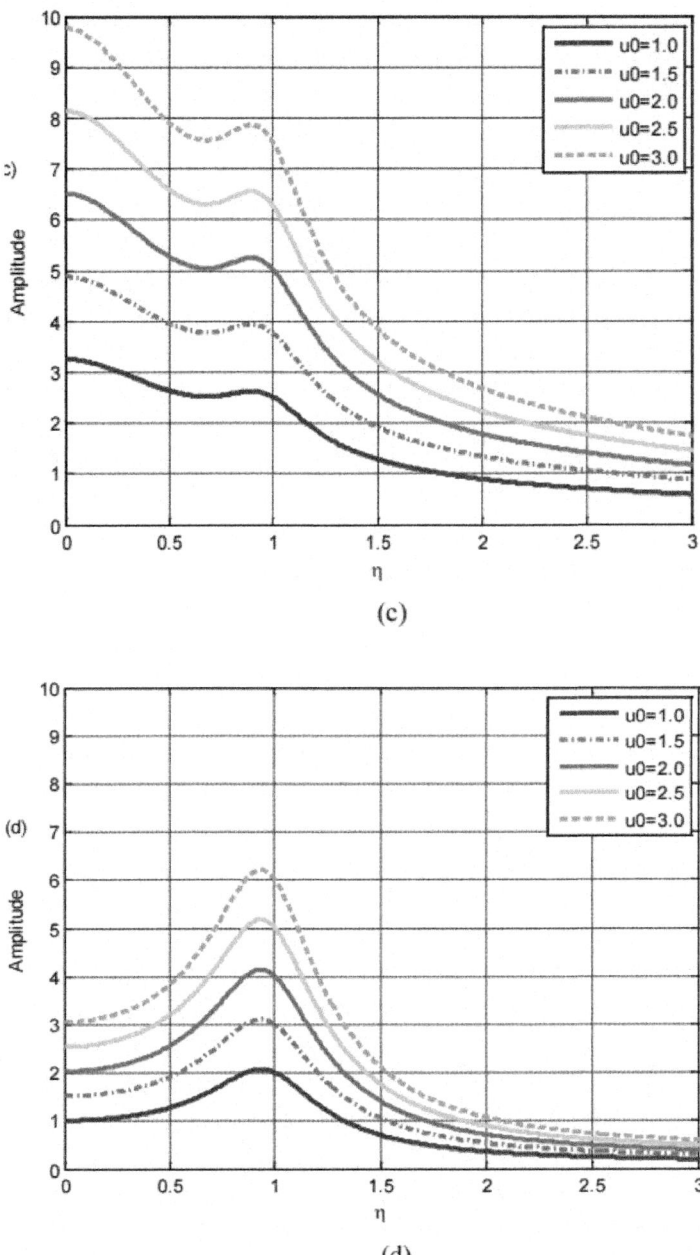

(c)

(d)

Figure 3: Amplitude as a function of frequency $\eta = \frac{w_1}{w_0}$ for various values of u_0 : (a) $\lambda = \frac{1}{4}$; $v = 1.15$; (b) $\lambda = \frac{1}{4}$; $v = 2.5$; (c) $\lambda = \frac{1}{2}$; $v = 1.15$; (d) $\lambda = \frac{1}{2}$; $v = 2.5$.

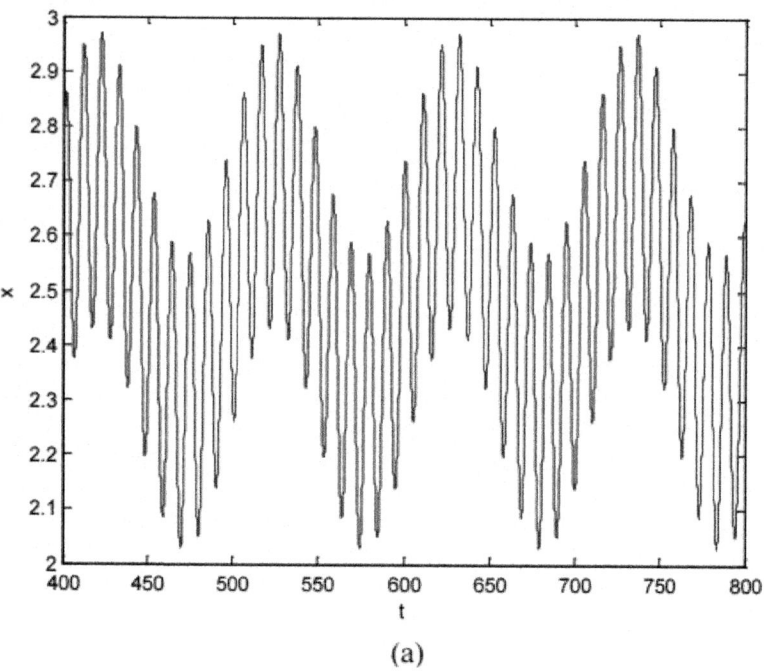

(a)

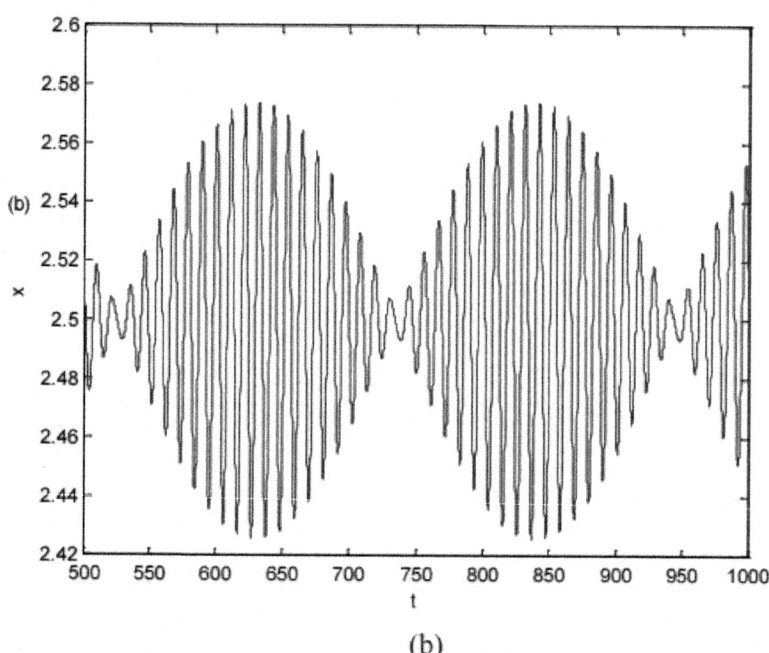

(b)

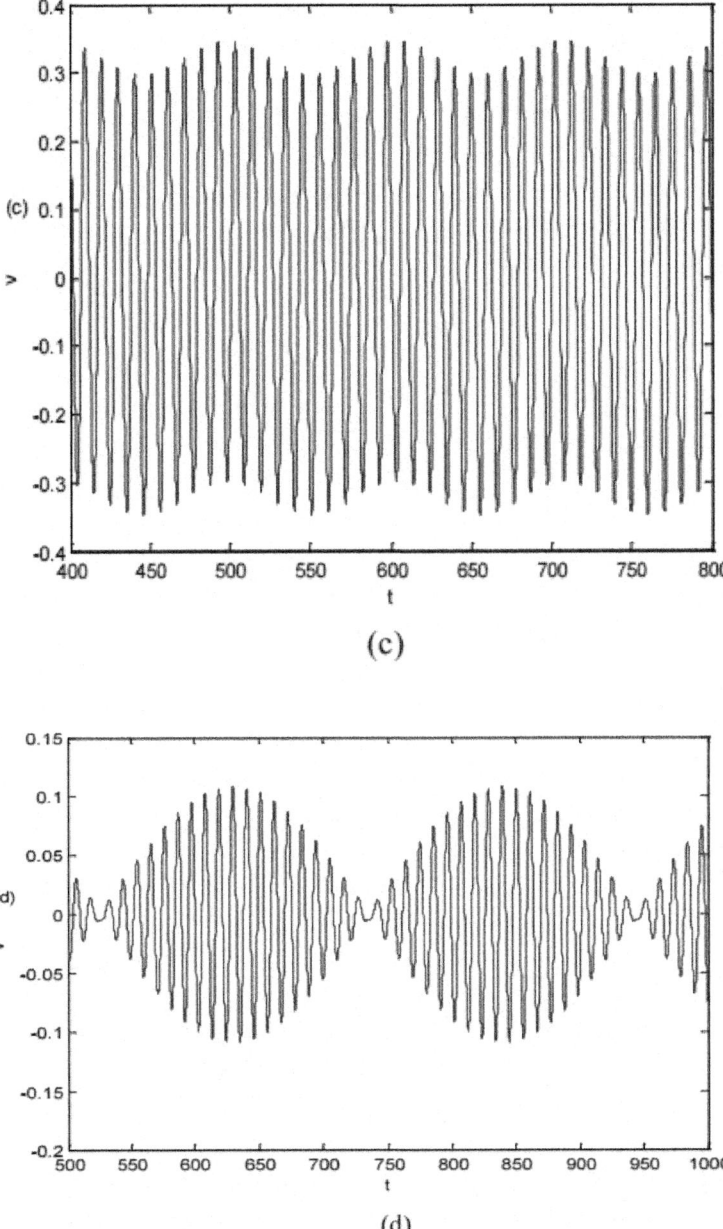

Figure 4: Modulated Motion without stick (top): (a) angular modulated: WM_1 ($n =$ 0.10; $u_0 = 0.2$); (b) Amplitude Modulated: WM_2 ($n = 0.95$; $u_0 = 0.05$) and corresponding velocity time history (bottom).

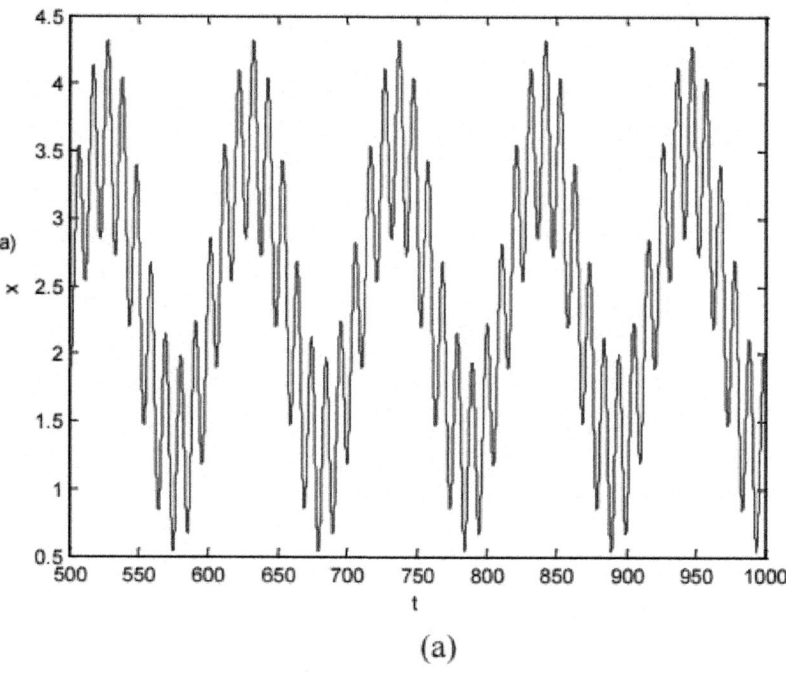

(a)

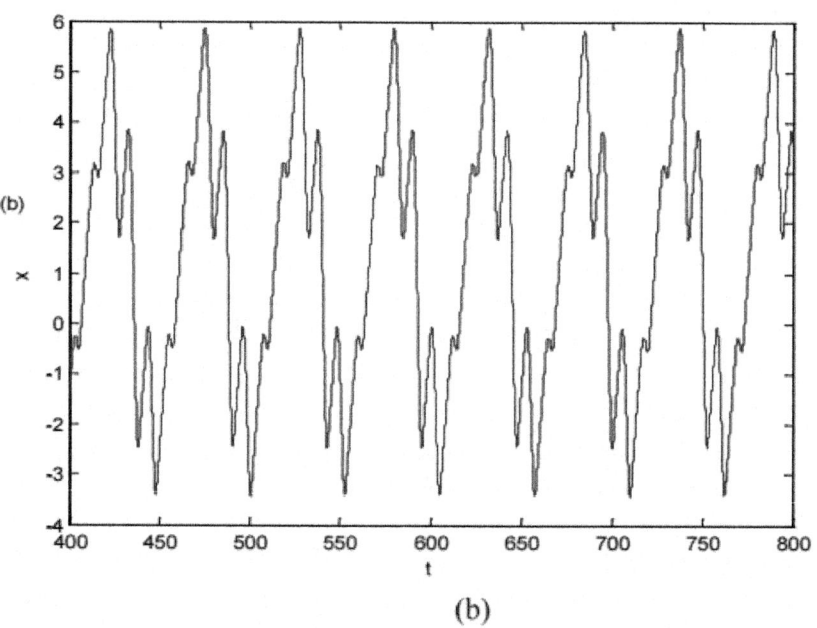

(b)

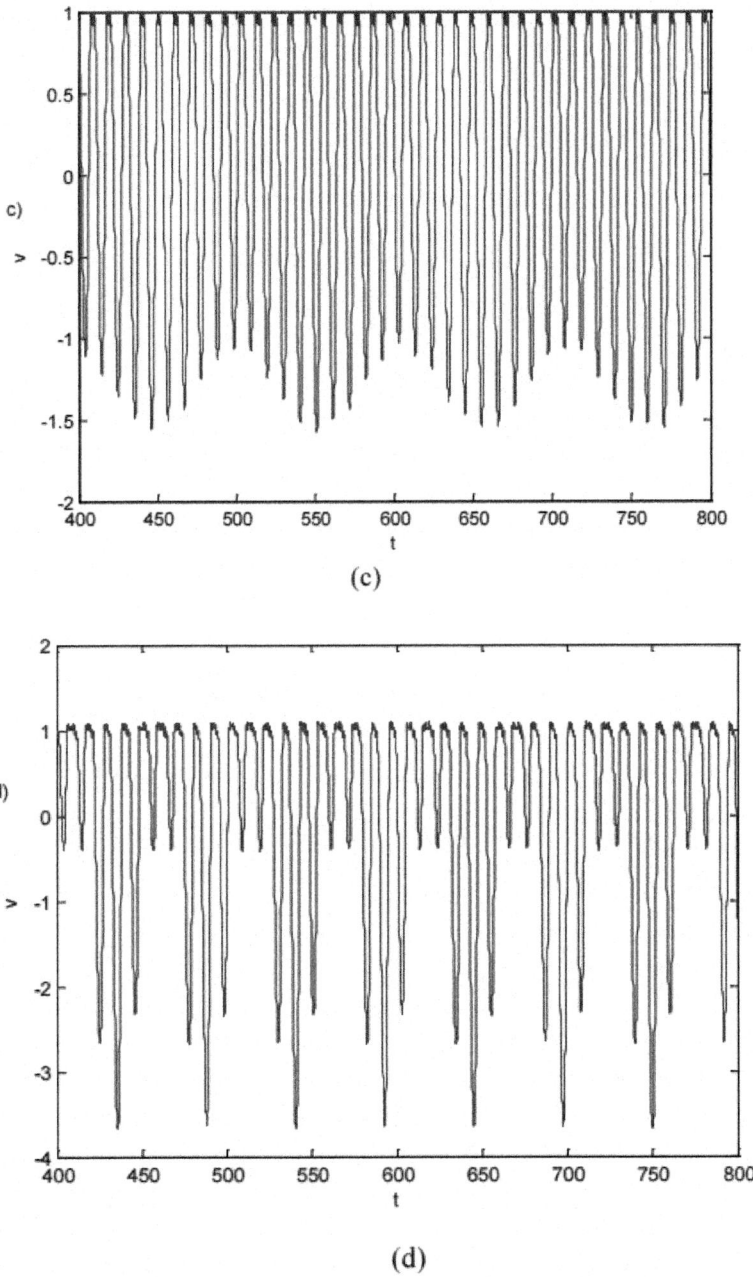

Figure 5: Modulated Motion with stick phase (SSM_1): (a) ($n = 0.10$; $u_0 = 1.15$); (b) $n = 0.2$; $u_0 = 3.05$) and corresponding velocity time history (bottom).

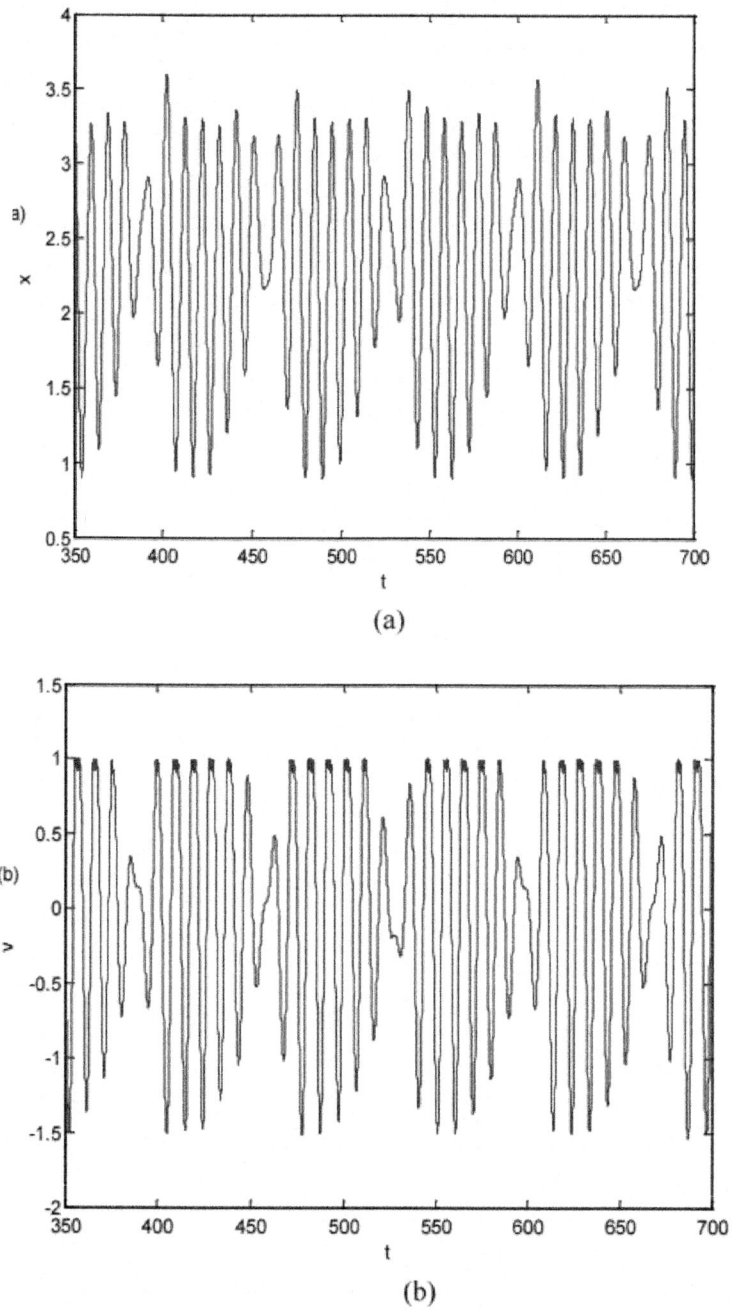

(a)

(b)

Figure 6: Modulated amplitude stick-slip (SSM_2); corresponding velocity time history ($n = 1.15$; $u_0 = 0.7$).

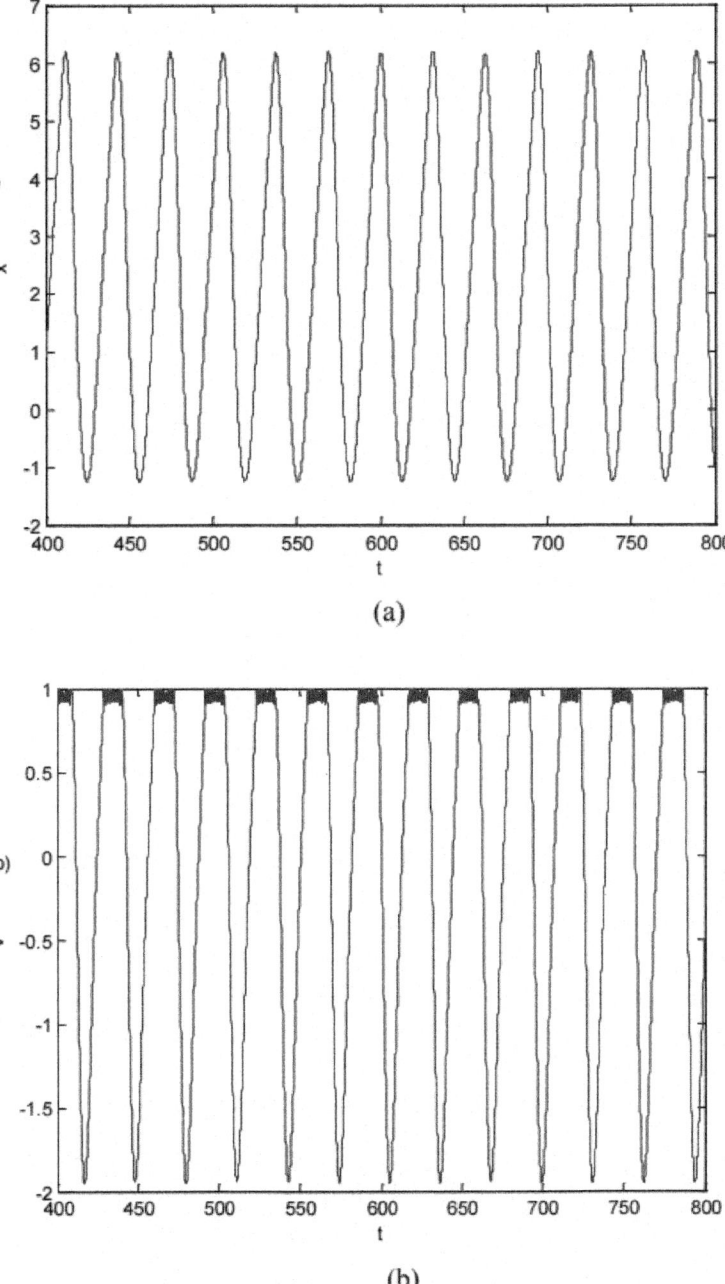

(a)

(b)

Figure 7: Periodic stick slip motion: (a) Periodic Oscillation: $\eta = 1/5$; $n = 1.0$; $u_0 = 1.65$; (b) Corresponding velocity time history.

a)   Wave modulated: we observed two types; wave modulated 1 (WM_1) which corresponds to the angular modulated wave and wave modulated 2 (WM_2) which refers to amplitude modulated. Figure 4 illustrates these forms of modulation with the corresponding time velocity obtained with the set parameter $v = 0.1$, $u_0 = 0.2$ for WM_1 and $v = 0.95$, $u_0 = 0.05$ for WM_2. The Block follows merely the dynamics of the excitation force $F_e(t)$. The two driving forces predominate the dynamics of the system and the effect of dry friction is not really observed.

b)   Stick-slip modulated: here the dynamics of the system presents modulated motion with stick phase. We observed two types regarding the form of the modulation of excitation force. Figure 5 presents stick-slip modulated 1 (SSM_1) which corresponds to the angular modulated stick-slip with the corresponding time velocity obtained with the set parameter n = 0.1, $u_0 = 1.15$ and n = 0.2, $u_0 = 3.05$. At the stick phase, the velocity remains constant and equal to the driving velocity $V_v = 1$, but the amplitudes of the block velocity during slipping are modulated. The stick phase is significantly presents here because of the effect of dry friction which predominated the dynamics of the system in comparison with the first case where the excitation forces influence significantly.

In Figure 6 there is no stick in all the motion cycle, but the velocity is modulated at slip phase. Figure 6 presents stick-slip modulated 2 (SSM_2) which corresponds to the amplitude stick-slip, this phenomenon occur for the set parameter n = 1.15, $u_0 = 0.7$.

c)   Periodic stick-slip and periodic oscillation: regarding the form of the excitation force, periodic stick slip occurs with integer values range of n. As n becomes an integer multiple, n, of the forcing ratio of the excitation frequency, the system has gone into periodic oscillation. The force degenerates to a single excitation case for n = 1. Figure 7 presents a periodic stick slip i.e., without any modulation and the corresponding time velocity obtained with the set parameter $\eta = 1/5$, $u_0 = 1.65$, n = 1.0.

Parameter's Map and Influence of Friction Laws

As an aid in describing and understanding nonlinear systems, someone can introduce maps in this section. In order to develop a meaningful understanding of friction experiments, and to predict dynamic system response and performance, an influence of the friction model must be studied. Although models predict well-defined stick-slip frequencies, intervals between successive stick-slip events have relatively broad distributions. For a better understanding of the map, the x-axis represents the amplitude of the applied u_0 response while, the y-axis represents the excitation frequency n. The first curve indicated the boundary between the Wave Modulated(WM) motion and the Stick-Slip Modulated (SSM), and the second curve indicated the limit between the end of (SSM) motion and the continue intermittent (WM). From the parameter maps, it is observed that specific motion lies on the special region of the parameters.

Figure 8 shows the dynamical phase diagram (in the n-u_0 plane), which presents regions of the ratio excitation frequency that correspond to different regimes of the motion of the block. Wave modulated 1 (WM_1) and

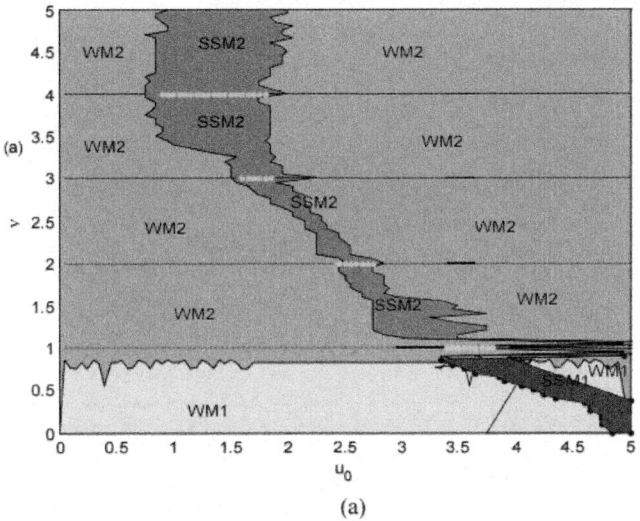

(a)

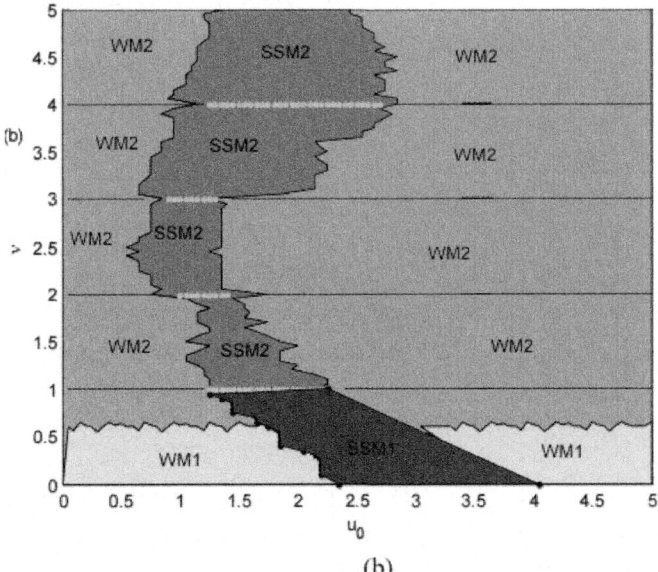

(b)

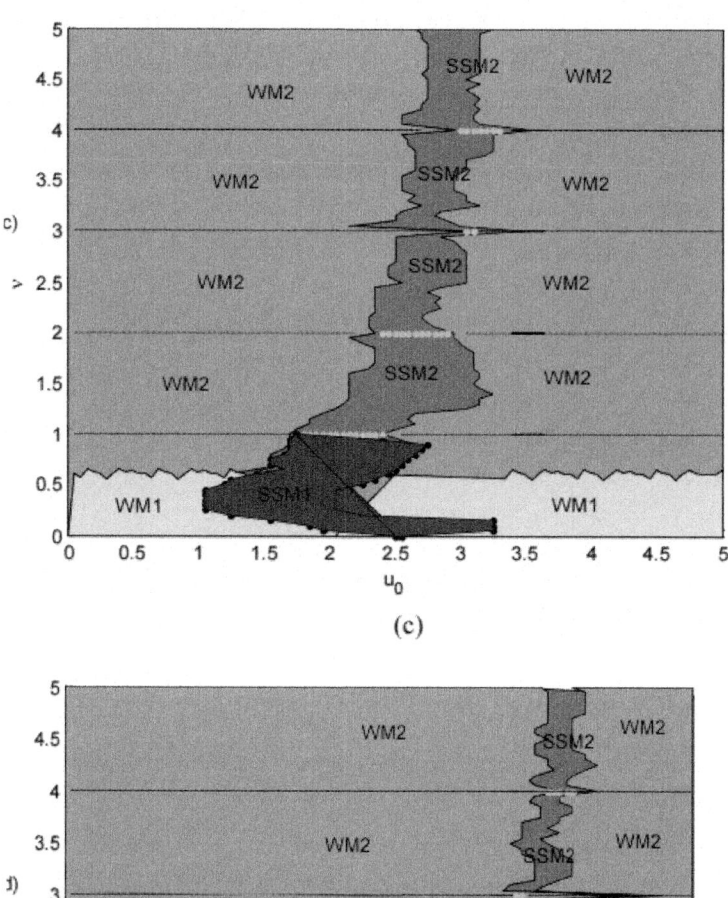

(c)

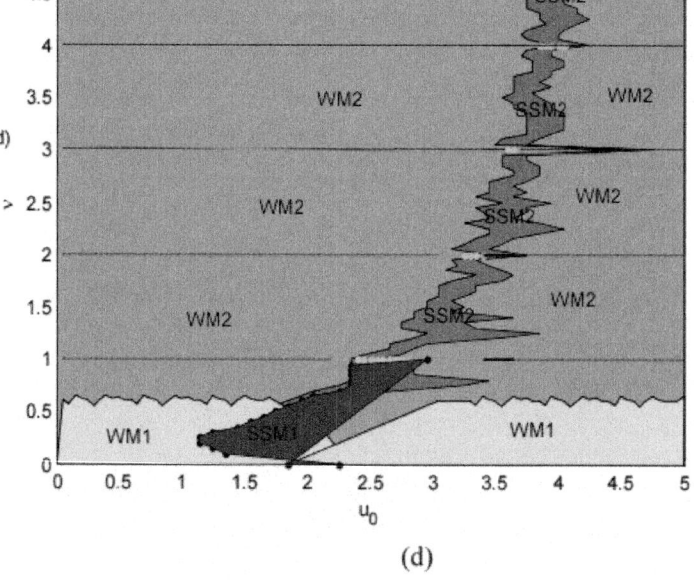

(d)

Figure 8: Parameters map of (u$_0$; n) using Coulomb law: (b$_1$ = 0.0; b$_3$ = 0.0; λ= 1/2; V$_v$ = 1): (a) η = 1/9; (b) η = 1/5; (c) η = 5/3; (d) η = 7/3.

Wave modulated 2 (WM_2) motion occur, respectively, to the left and right at the first and second line (with n < 1) and to the left-right of the first-second line (with n > 1). The system exhibits stick-slip modulated (SSM) motion in the range of parameters between these two curves. The two lines describe the n dependence of the applied u_0 corresponding to the transitions between different states of motion. The parameter η gives the orientation of these two curves. For instance, in Figure 8(a) and Figure 8(b), η is very weak, respectively 0.11 and 0.2, that make the slop of limits curve lying to the left side of the diagram close to n-axis apses, this prove that the applied u_0 decreases significantly. The important observation is that for n comprises in between 1 and 3, the value of stick-slip modulated (SSM_2) is mostly nil in Figure 8(a), [respectively nil in Figure 8(b)], and some values of u_0 are not inside the diagram for certain values surrounding the frequency n = 1. When n out dated the value 3 to u_0 weak, the stick-slip bandwidth is large.

As an interesting details, one can observe that if w_1 smaller than the natural frequency w_0, the bandwidth became bigger as n increase and for weakest values of η, the stick-slip modulated begin for very large values of u_0, the response to a periodic oscillation is described in all figures in parameter maps for integer ratio n outcast stick-slip band. The bright reflecting type points indicated the stick-slip modulated inside stick-slip band and the dash one represent the periodic stick-slip. For larger values of the excitation amplitude depending on the excitation frequency, the friction oscillator shows periodic motions modulated.

When n is very weak, we have u_0 very large in Figure 8(b), we observed the presence of periodic stick-slip in between the stick-slip bandwidth for the integer value of n while the periodic oscillation in two extreme of always integer value of n. We also observe that the stick-slip modulated (SSM_1) ends for n = 1 while stick-slip modulated (SSM_2) begins from n > 1. Figure 8(c) and Figure 8(d) illustrated the case where η > 1 and the two curves are lying to the right of the diagram. As the friction force increases, the stick motion is suppressed in Figure 8(d), (η = 2.33) [respectively in Figure 8(c) (η = 1.67)]. Note that the stick-slip motion tends to disappear if w_1 is bigger than w_0 in the ratio η so that the big values of η diminish significantly the bandwidth of stick-slip. Using this parameter map the system behaviour can be characterized for any set of parameters within the plotted range.

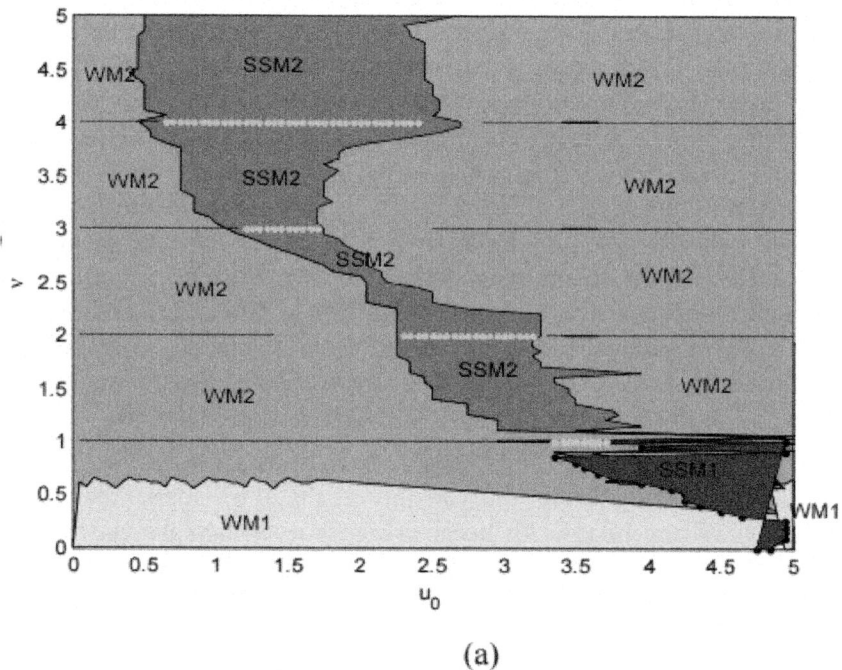

(a)

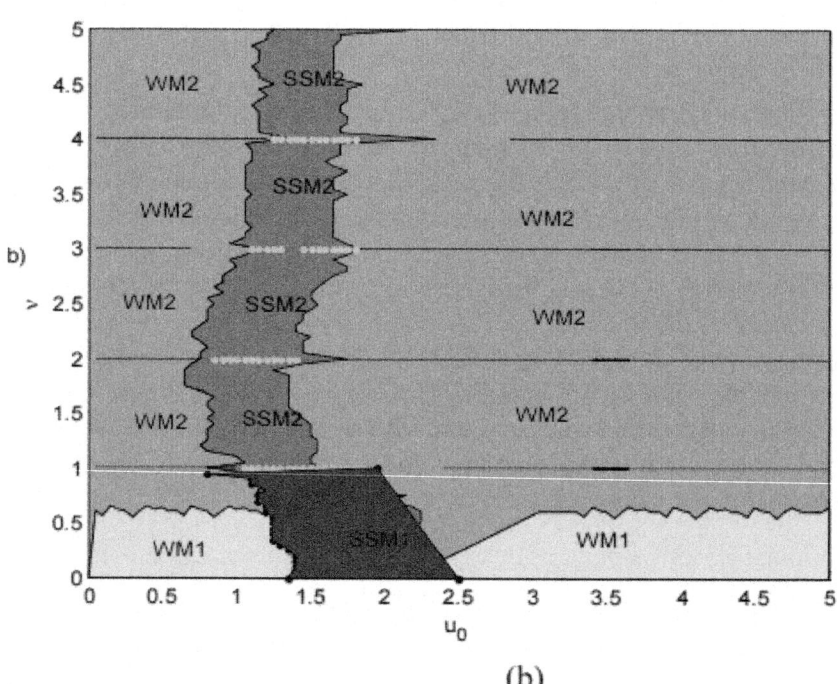

(b)

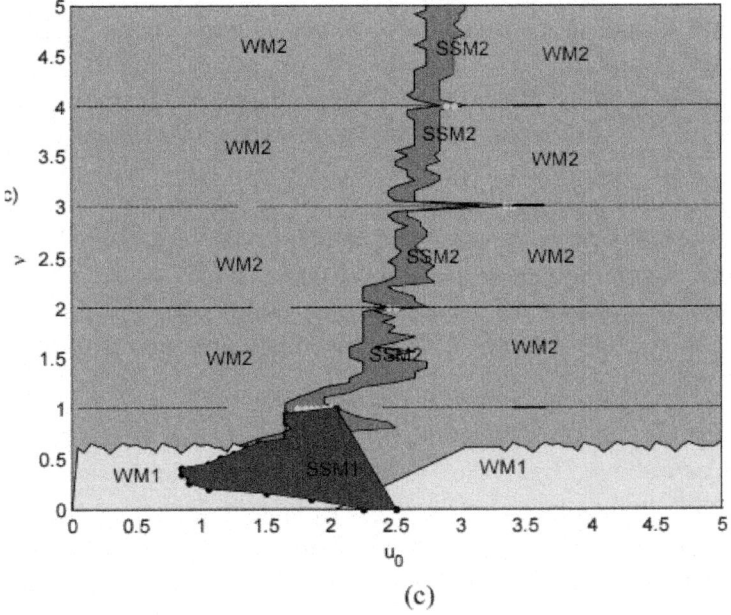

(c)

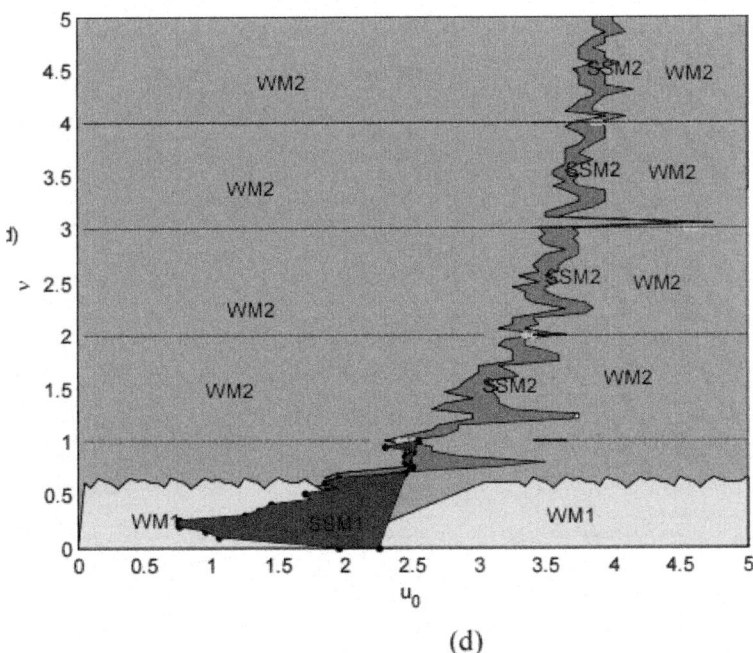

(d)

Figure 9: Parameters map of $(u_0; n)$ using Coulomb + viscosity: $(b_1 = 0.2; b_3 = 0.0; \lambda = 1/2; V_v = 1)$: (a) $\eta = 1/9$; (b) $\eta = 5/17$; (c) $\eta = 5/3$; (d) $\eta = 7/3$.

Using the same parameters system as in Figure 8, we reproduce the simulation with the two other friction forces mention above. In the case of the friction with Coulomb + viscosity, ($b_1 = 0.2$; $b_3 = 0.0$), Figure 9 shows the same observations as above, but the SSM_1 is mostly nil or outside the $n - u_0$ plane. If $n \in [2]$ [3] the bandwidth SSM tend to disappear (very weak). For $\eta = 0.29$, the two limits curves are relatively straight in Figure 9(b). But Figure 9(c) and Figure 9(d) illustrated the curve with w_1 greater than w_0, the amplitude u_0 takes very large values, the stick-slip band become very weak. In the mapping procedure, special attention has to be paid to the lower or the higher values of the ratio η. we observed a diminishingly amplitudes u_0 if the frequency w_1 is less than w_0. The features of high amplitude and high n-factor are clearly recognized.

In the case of Stribeck friction, $b_1 \neq 0$; $b_3 \neq 0$, Figure 10 is not very different from the case above.Figure 10(c) with $\eta = 0.6$ gives a relative straight limit curve of stick-slip modulated (SSM) and for w_1 greater than w_0, another parameter map showing the influence of the excitation amplitude u_0 is represented and the stick-slip phenomenon disappear partially in Figure 10(d).

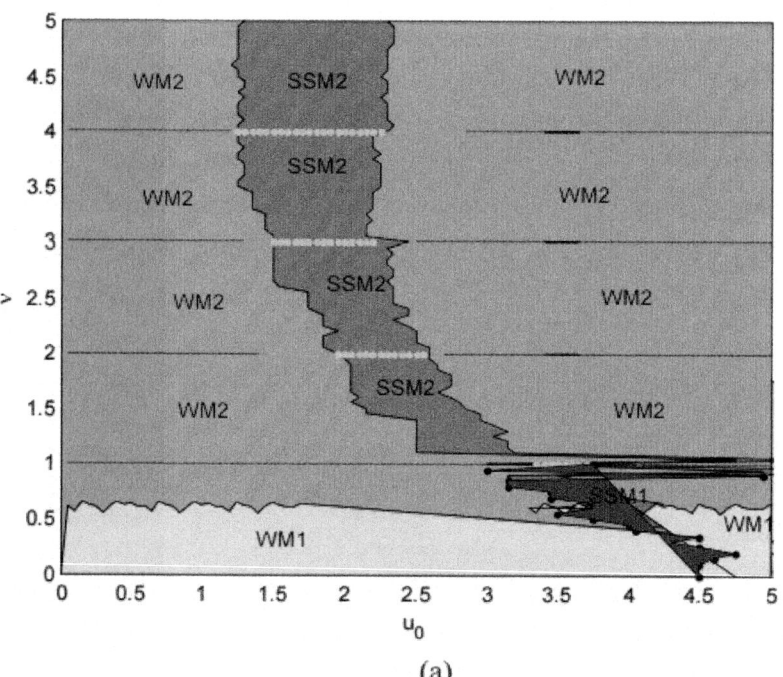

(a)

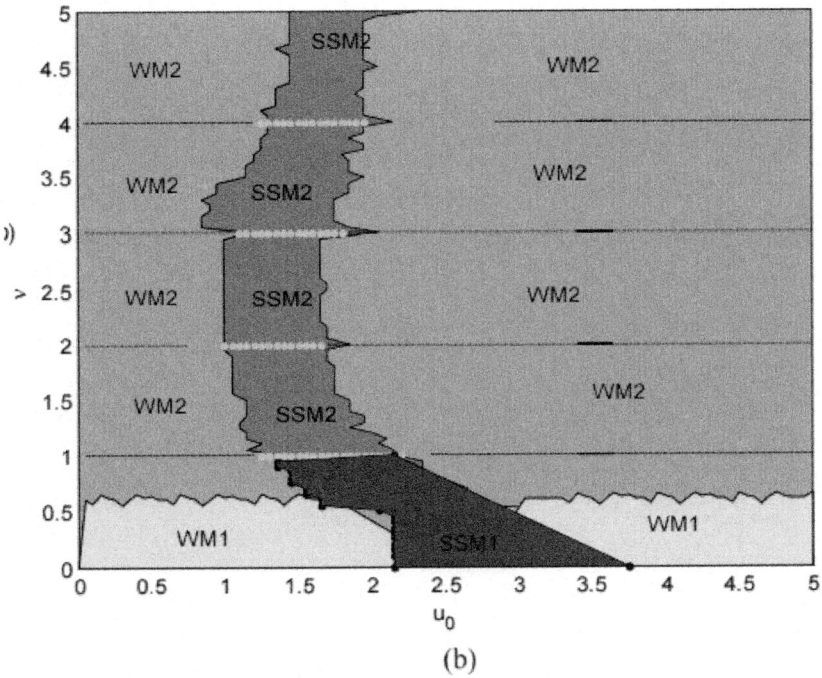

(b)

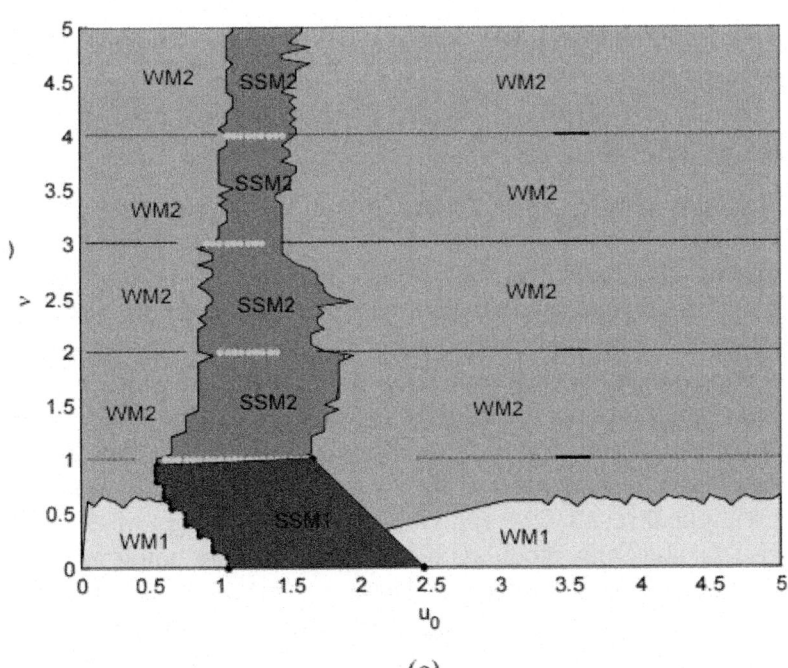

(c)

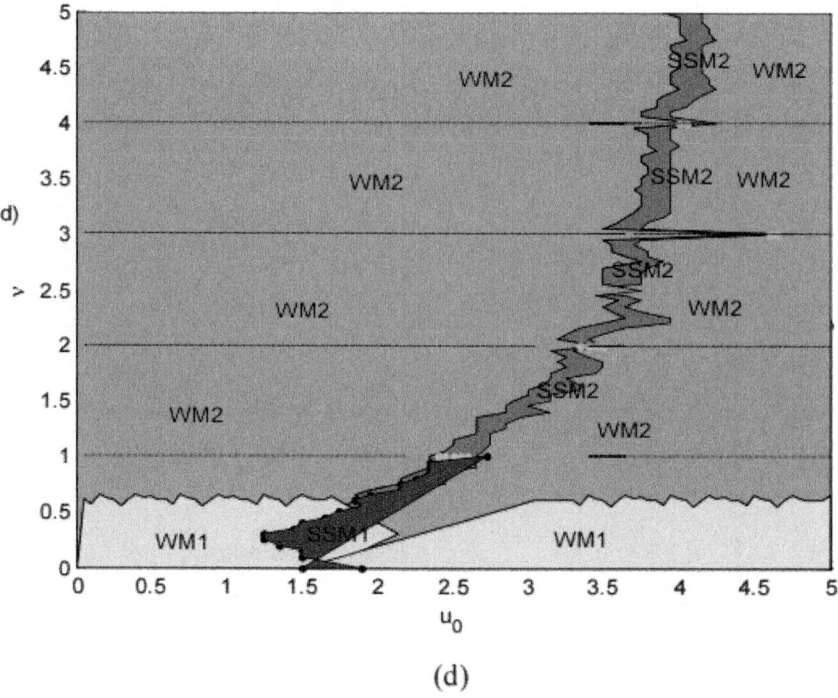

(d)

Figure 10: Parameters map of $(u_0; \eta)$ using Stribeck friction laws: $(b_1 = 0.2; b_3 = 0.16;$ $\lambda = 1/2; V_v = 1)$: (a) $\eta = 1/9$; (b) $\eta = 1/5$; (c) $\eta = 3/5$; (d) $\eta = 7/3$.

CONCLUSION

The first main purpose of this paper was to investigate the influence of the frequencies, in external excitation with two harmonic driving forces in the case of a moving base. The second main purpose was to show the influence of various friction characteristics and parameter maps on the system response. Numerical and analytical predictions of the scenarios varying with the ratios of the two excitation frequencies and amplitudes are carried out, and the parameter maps for specific motions are presented. This study suggests that, the transition between each of the motions strongly depends on the parameters (frequencies and amplitudes) of the two driving forces; this was appreciated in the analytical result. The applying two driving forces were also found to give rise to stick-slip modulation. We observed the amplitude modulation of the velocity during slip phase. These and other results contribute to the general understanding of how friction properties may change under the action of the two vibrating forces. If two harmonics of almost periodic excitation are interacted, modulated motions take place. But we must control the ratio η to control the dynamic of the system. In all mechanical system, Stick-Slip motion is considered as harmful effects. So that too many researches are concentrated to diminish amplitude of vibration which can directly affect stresses and thus the life of the system and our results have potentially an equally wide range of applications in engineering.

REFERENCES

1. Ewins, D.J. (2001) Modal Testing: Theory, Practice and Application, Mechanical Engineering Research, Studies Engineering Design Series. Research Studies Pre, 2nd Edition.

2. Piyawat, K. and Pei, J.S. (2005) Idealized Excitation Forces for Nonlinear SDOF Systems with Memory and Degradation. School of Civil Engineering and Environmental Science, The University of Oklahoma, Norman.

3. Den Hartog, J.P. (1931) Forced Vibrations with Combined Coulomb and Viscous Friction. Trans ASME, 53, 107-115.

4. Wojewoda, J., Kapitaniak, T., Barron, R. and Brindley, J. (1993) Complex Behaviour of a Quasiperiodically Forced System with Dry Friction. Chaos, Solitons & Fractals, 3, 35-46.http://dx.doi.org/10.1016/0960-0779(93)90038-3

5. Shaw, S.W. (1986) On the Dynamic Response of a System with Dry Friction. Journal of Sound and Vibration, 108, 305-325. http://dx.doi.org/10.1016/S0022-460X(86)80058-X

6. Ibrahim, R. (1994) Friction Induced Vibration, Chatter, Squeal and Chaos—Part I: Mechanics of Contact and Friction. Applied Mechanics Review, 47, 209-226.http://dx.doi.org/10.1115/1.3111079

7. Andreaus, U. and Casini, P. (2001) Dynamics of Friction Oscillators Excited by a Moving Base and/or Driving Force. Journal of Sound and Vibration, 245, 685-699.http://dx.doi.org/10.1006/jsvi.2000.3555

8. Stefannski, A., Wojewoda, J., Wiercigroch, M. and Kapitaniak, T. (2003) Chaos Caused by Non-Reversible Dry Friction. Chaos, Solitons and Fractals, 16, 661-664.http://dx.doi.org/10.1016/S0960-0779(02)00451-4

9. Wiercigroch, M., Vwt, S. and Zfk, L. (1999) Non-Reversible dry Friction Oscillator: Design and Measurements. Proceedings of the Institution of Mechanical Engineers, 213, 527-534.

10. Fenny, B. and Moon, F.C. (1994) Chaos in a Forced Dry Friction Oscillator: Experiment and Numerical Modelling. Journal of Sound and Vibration, 170, 303-323.http://dx.doi.org/10.1006/jsvi.1994.1065

11. Kardan, I., Kabganian, M., Abiri, R. and Bagheri, M. (2013) Stick-Slipconditions in the General Motion of a Planar Rigid Body. Journal of Mechanical Science and Technology, 27, 2577-2583.

12. Cao, Y. and Chen, X.B. (2015) An ARX-Based PID-Sliding Mode Control on Velocity Tracking Control of a Stick- Slip Piezoelectric-Driven Actuator. Modern Mechanical Engineering, 5, 10-19. http://dx.doi.org/10.4236/mme.2015.51002

13. Thomsen, J.J. (1999) Using Fast Vibrations to Quench Friction-Induced Oscillations. Journal of Sound and Vibration, 228, 1079-1102.http://dx.doi.org/10.1006/jsvi.1999.2460

14. Cheng, G. and Zu, J.W. (2004) Dynamics of a Dry Friction Oscillator under Two-Frequency Excitations. Journal of Sound and Vibration, 275, 591-603.http://dx.doi.org/10.1016/j.jsv.2003.06.027

15. Awrejcewics, J. and Dzyubak, L. (2003) Stick-Slip Chaotic Oscillations in Aquasi-Autonomous Mechanical System. International Journal of Nonlinear Sciences and Numerical Simulation, 4, 155-160.

16. Awrejcewicz, J. and Pyryev, Yu. (2002) Thermo Elastic Contact of a Rotating Shaft with a Rigid Bush in Conditions of Bush Wear and Stick-Slip Movements. International Journal of Engineering Science, 40, 1113-1130.

17. Voldrich, J. (2009) Modelling of Three-Dimensional Friction Contact of Vibrating Elastic Bodies with Rough Surfaces. Applied and Computational Mechanics, 3, 241-252.

18. Reinhorn, A.M. and Sivaselvan, M.V. (2000) Hysteretic Models for Deteriorating Inelastic Structures. ASCE Journal of Engineering Mechanics, 126, 633-640.http://dx.doi.org/10.1061/(ASCE)0733-9399(2000)126:6(633)

19. Sergienko, O.V., Macayeal, D.R. and Bindschadler, R.A. (2009) Stick-Slip Behavior of Ice Streams: Modelling Investigations. Annals of Glaciology, 50, 87-94.

Chapter 7

DEVELOPMENT OF A MULTIPLE DEGREE OF FREEDOM KNEE DISARTICULATION PROSTHESIS WITH ACTIVE LEG LENGTH VARIATION

Berend Denkena, Martin Eckl, and Dominik Brouwer

Leibniz Universität Hannover, Institute of Production Engineering and Machine Tools, An der Universität 2, Hanover, Germany

ABSTRACT

This paper presents a novel design for a knee disarticulation prosthesis. In this design, three hydraulic cylinders form the supporting structure and provide the damping effect at the same time. That way the novel knee joint offers two fundamental advantages compared to the state of the art. First, the combination of a supporting structure and damping element reduces the weight of the prosthesis. Secondly, the use of several cylinders allows the actuation of further degrees of freedom. Additional degrees of freedom can be used to vary the leg length within the gait cycle and hence to optimize the gait behavior.

INTRODUCTION

Exoprostheses are orthotics to balance functionality and cosmetic aspects of lost extremities. In the past 100 years the limb amputation indication changed from life saving measures due to injury to plastic, micro surgical intervention. The indication and the individual height of amputation influence the rehabilitation course. Hagberg et al. show that the higher the level of amputation, the lower the degree of rehabilitation; hence a below knee amputation should be aspired [1] . Figure 1 shows the different amputation heights.

If a higher level of amputation is not avoidable, knee disarticulation is to be considered rather than above knee amputation. Knee disarticulation was the amputation through the knee joint and was performed first in 1825 in the USA [2] . Compared to above knee amputation, knee disarticulation offers many advantages. The knee stump keeps the ability to absorb the weight force as the ischial tuberosity is spared [3] . Furthermore, the knee disarticulation does not affect the femur, and patella [4] . Thus, the remaining thigh muscles allow for coordinated stump movement [5] . In case of knee disarticulation, many different designs of the artificial knee joint can be applied. Stark gives a brief overview of such different designs [6] . In principle, any prosthesis for above knee amputees is suitable for knee disarticulation amputees as well. However, as the natural knee stump is still present, the artificial knee joint would be placed below the natural knee. These different anatomic situations need to be considered when selecting a knee joint design.

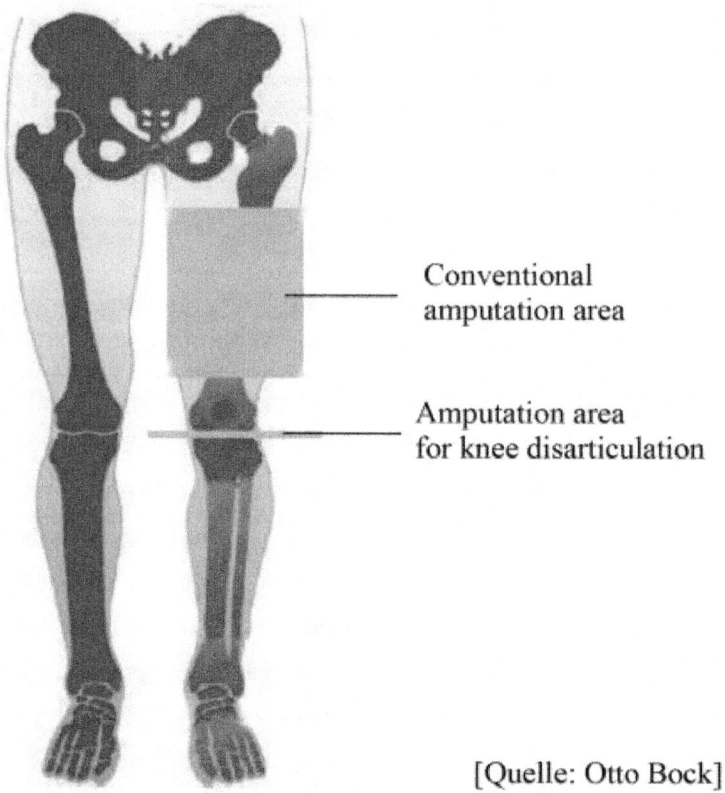

Conventional
amputation area

Amputation area
for knee disarticulation

[Quelle: Otto Bock]

Figure 1: Schematic representation of the different amputation levels.

Current artificial knee joints are designed as single axis or polycentric joints. When single axis joints are used for knee disarticulation amputees, the distance between the anatomic and the artificial knee pivot point is long. This leads to an unnatural gait behavior and less stability within the walking cycle. O'Connor introduces a single axis joint where a special knee bracket reduces the distance between the anatomic knee and the pivot of the artificial knee and provides a more naturally looking and operating prosthetic leg [7] . In contrast to single axis joints, polycentric knee joints are mainly used for knee disarticulation amputees. In this case, the pivot is variable and outside of the actual knee joint mechanism. The design of such polycentric knee joints can be optimized to the requirements of knee disarticulation amputees. The University of Toledo has developed a method to optimize a polycentric knee joint for knee disarticulation prostheses [8] . All current knee joints consist of a mechanical supporting structure and an additional damping element.

In contrast, this paper presents a novel design for a polycentric knee joint for knee disarticulation prostheses, where hydraulic cylinders form the supporting structure and provide the damping effect at the same time (Figure 2).

The novel knee joint offers two fundamental advantages compared to the state of the art. First, the combination of a supporting structure and damping element reduces the weight of the prosthesis. Secondly, the use of several cylinders allows the actuation of further degrees of freedom. Additional degrees of freedom can be used to vary the leg length within the gait cycle and hence to optimize the gait behavior. The patented design [9] was developed within the cooperative research project MultiPro, funded by the Federal Ministry for Research and Education (BMBF).

This paper is organized as follows. Section 2 explains the design and the functionality of the named knee joint mechanism, followed by the control scheme. Further, a test rig is presented, which is able to simulate a gait cycle. Section 3 discusses the mathematical relations between the mechanical and the hydraulic part of the knee joint. Based on these mathematical relations, a control scheme is developed to absorb the torque, generated by the weight of the patient and the lever arm between the line of action of the weight force and the prosthesis. Furthermore, a simulation model of the mechanism is developed using the software tools SimMechanics and SimHydraulics (Mathworks). In Section 4 the hydraulic scheme is extended to enable active leg length variation within the gait cycle. The control scheme is simulated using Matlab/Simulink. That enables the simulation of a whole gait cycle with leg length variation. Finally, the development results of the research project are summarized in the conclusion.

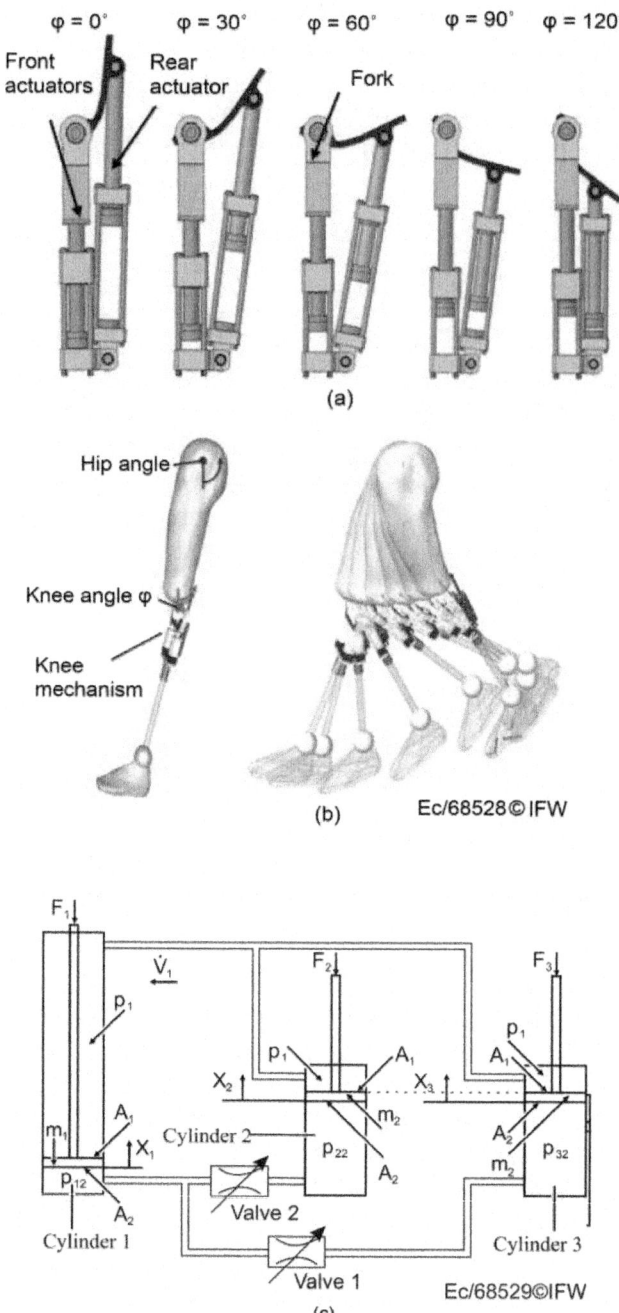

Figure 2: Design of the knee mechanism. (a) CAD model for different knee angles; (b) Knee mechanism within a gait cycle; (c) Hydraulic system overview.

DESIGN OF THE KNEE MECHANISM

Description of the Novel Knee Mechanism

Figure 2(a) shows the CAD model of the artificial knee mechanism for different knee angles. The knee mechanism consists generally of three linear hydraulic actuators. Two of the actuators are arranged symmetrically in the front of the prosthesis, while the third is arranged at the back of the prosthesis. Figure 2(b) illustrates the novel knee mechanism and a leg within the gait cycle (the kinematics of a gait cycle is shown in Figure 2). The left-top part of Figure 2 depicts how different angles are realized with the novel knee mechanism. To realize the knee of zero degrees, the front cylinders are at zero stroke and the back cylinder is at maximum stroke. To realize an angle, the forward cylinders are extended, the rearward cylinder is retracted. In this way, angles up to 120° can be realized. If the front cylinders and the rear cylinder are retracted, the length of the prosthesis is shortened, if all actuators are extended, the prosthesis is extended accordingly.

Figure 2(c) shows the hydraulic circuit diagram. The linear actuators are realized as hydraulic cylinders in a closed system. The volume of two front cylinders (indices 2 and 3) is equal to the volume of the rear cylinder (index 1). In this way, only the desired defined knee-angles can be realized. To implement a damping functionality, two electronically controllable throttle valves are integrated. To further implement the functionality of controlled leg length change, additional components have to be included. These components are described in Section 4. The change of the knee angle is driven by the external influences: hip movement, the patient weight force, and the ground reaction force. In the swing phase, the hip is moved forward and, thus, because of the inertia of the prosthesis, the prosthesis leg. In the stance phase the foot is fixed at the ground and the body is moved forward mainly by the hip movement. The ground reaction force and the movement of the hip generate a torque at the fork. So, in both phases of the gait, a counter direction movement of the rearward cylinder (cylinder 1) and the forward cylinders (cylinders 2 and 3) is generated. The fluid flows through the proportional valves 1 and 2 and the knee angle can be damped by adjusting the cross section of the valves to generate a sufficient torque in the stance phase and to control the correct knee angle in the swing phase. The damping value is set by proportional valves, depending on the current gait phase.

The invented knee mechanism is designed to sustain the mechanical stress, required by standard ISO10328 [10] . Most parts have been optimized using the Finite Element Method. If possible, lightweight materials have been used. The parts of the prosthesis, their materials and the corresponding mass are shown in Table 1. The knee mechanism has a weight of 1.115 kg while the whole prosthesis weighs 2.5 kg and is designed for patients with a maximum weight of 125 kg. A maximum knee angle of 136° is attainable.

Table 1: Overview of the components and its weights used in the knee joint.

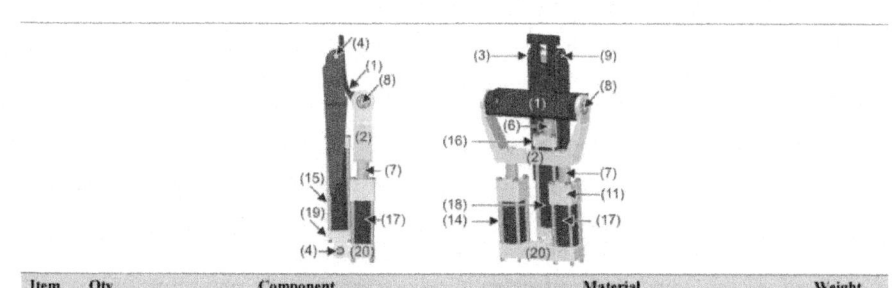

Item	Qty	Component	Material	Weight
1	1	Socket adapter	CFRP	78 g
2	1	Suspension fork	Aluminum	132 g
3	1	Brace	CFRP	72 g
4	2	Joint bolt (rearward)	Stainless Steel	16 g
5	4	Bearing bushing (8 mm)	Iglidur-W300	0.5 g
6	1	Piston and piston rod (rearward)	Stainless Steel	99 g
7	2	Piston and piston rod (forward)	Stainless Steel	109 g
8	2	Joint bolt (forward)	Stainless Steel	10 g
9	2	Screw M6×25 (DIN 7984)	Stainless Steel	8 g
10	2	Bearing bushing (15 mm)	Iglidur-W300	0.5 g
11	2	Cylinder cover (forward)	Aluminum	33 g
12	2	Bearing bushing (13 mm)	Iglidur-W300	0.5 g
13	6	Piston end stop	PV	1 g
14	8	Tension rod (forward)	Stainless Steel	10 g
15	4	Tension rod (rearward)	Stainless Steel	12 g
16	1	Cylinder cover (rearward)	Aluminum	17 g
17	2	Cylinder tube (forward)	CFRP	26 g
18	1	Cylinder tube (rearward)	CFRP	44 g
19	1	Cylinder bottom (rearward)	Aluminum	23 g
20	1	Cylinder bottom (forward)	Aluminum	108 g
		Whole mechanism		1115 g

Controlling a Gait Cycle

The required damping value depends on the gait phase. During the stance phase, the knee has to provide enough resistance to generate a sufficient torque to accommodate the weight force of the patient. While in the swing phase, the damping has to be low but adaptive to generate a natural looking swinging movement. Thus, controlling a gait cycle requires the knowledge of the ground reaction forces in the stance phase, the current angle of the knee in the swing phase, and the desired movement of the knee. Any patient has his own pattern of movement, depending on the age, gender, body dimensions and, body proportions. Nevertheless, the general course of movement is nearly equal for any patient [11] [12] and called gait cycle. The kinematics of the knee and the hip joint within the gait cycle are depicted in Figure 3.

The shown cycle begins with the stance phase. In this phase, the hip is at around thirty degrees, the knee is strait, and the foot touches the ground. While the foot is touching the ground, the hip is rotated back to approx. zero degrees and the knee is bended to approx. eighteen degrees. In this phase, the body is moved forward. In the swing phase of the illustrated leg (the other leg would conduct the stance phase), the hip is rotated to the starting position, while the knee has to be rotated to around sixty degrees to avoid collision of the toes with the ground.

The gait cycle begins with the stance phase while the foot is in contact with the ground. Within the swing phase, however, there is no contact between the foot and the ground. The lower part ofFigure 4 shows the horizontal (F_H) and vertical (F_V) ground reaction force, based on the average measuring results, presented in [12] -[18] . The upper part (right) shows a model of the location of the ground reaction force on the foot depending on the gait cycle which was used for the simulations in this contribution. The point of application of the force vector depends on the gait phase. To be able to implement the ground reaction force in the simulation model, it is modeled as a superposition of the three vectors at the points of application (a_1-a_3, Figure 4(b). The according vectors are weighted by the time dependent variables $a_1(t)$, $a_2(t)$, and $a_3(t)$. The sum of these variables is 1 at any time:

$$F_B(t) = a_1(t) \cdot F_B(t) + a_2(t) \cdot F_B(t) + a_3(t) \cdot F_B(t) = F_H \cdot e_x + F_V \cdot e_y \qquad (1)$$

At the beginning of the stance phase the heel has the first contact with the ground: a_1 is 1, a_2 and a_3 are equal to zero. Within the stance phase, the point of application moves to the toes. In this way, at the end of the stance phase, a_3 is 1 and a_1 and a_2 are equal to zero (Figure 4(a)).

With the knowledge of the external influences within a gait cycle a control scheme is developed (Figure 5). The knee mechanism, presented in this paper, damps the leg movement hydraulically using proportional valves. As a control algorithm a fuzzy controller with 25 fuzzy rules is used to calculate the optimal damping value D_φ depending on the gait phase and optimized by the optimization algorithm "active set" [19] [20] . The input values of the fuzzy controller are the pressures inside the knee mechanism (P_1 and P_2), as well as the knee angle φ. It is necessary to convert the necessary damping value D_φ from the controller to the resulting damping values of the pistons. For this purpose, the direct kinematic is defined analytically according to the underlying geometric relations. The damping of the pistons is realized by adjusting the values of the hydraulic resistances (R_1 and R_2) of the valves 1 and 2.

The calculation of the knee angle φ from the measured piston positions is based on the inverse kinematic, which is, however, very complex to deduce analytically. Instead, discrete values of the inverse kinematic are identified using the CAD Software ProEngineer© and approximated by a fourth order Fourier transformation [21] . The calculation of the cross section of the proportional valves A_v from the damping values of the pistons is derived in [22] .

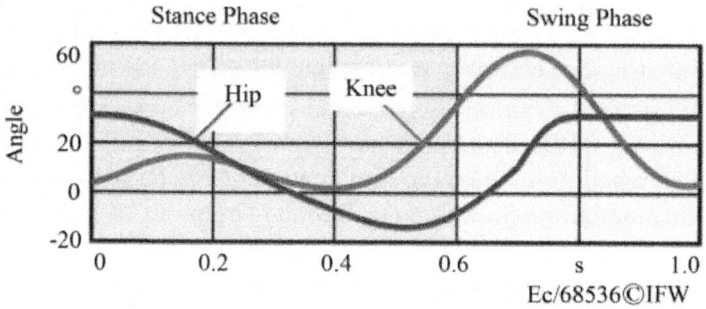

Figure 3: Pattern of movement. (a) Curve of the knee angle; (b) Curve of the hip angle.

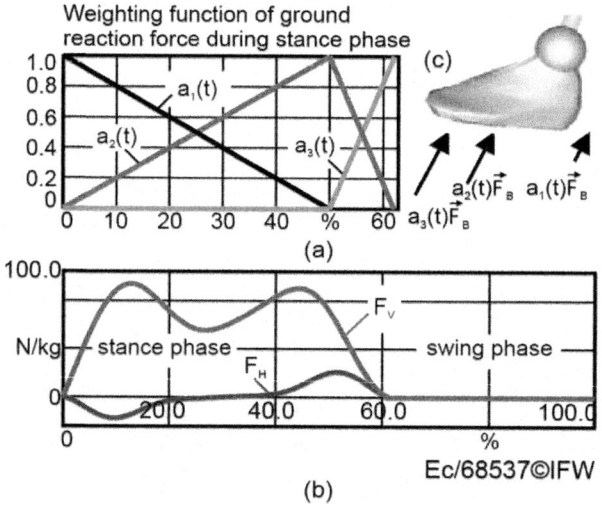

Figure 4: Kinetic influence of the walking cycle. (a) Weight- ing of the points of action; (b) Locations of the points of action; (c) Force curves within one gait cycle.

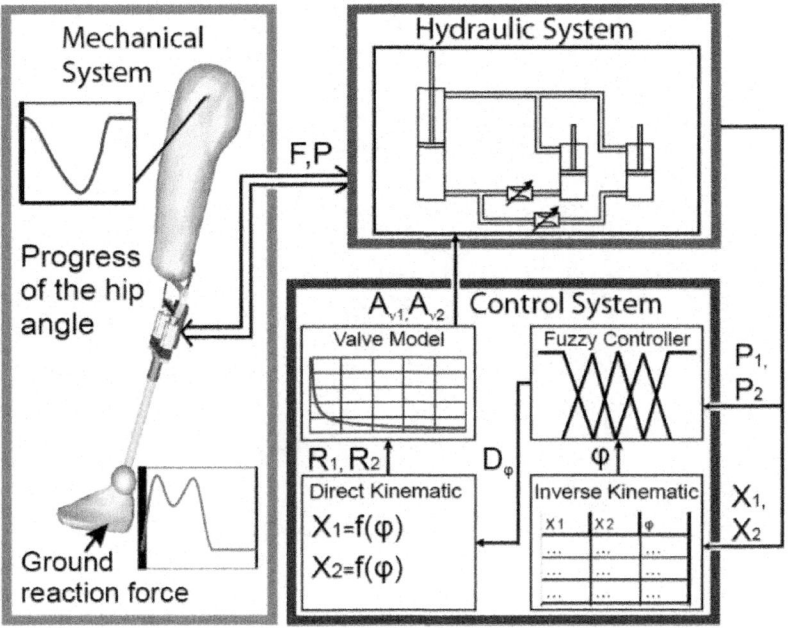

Figure 5: Control loop of the knee mechanism.

$$A_v = \sqrt[3]{\frac{\pi \rho^2 \mathrm{Re}_{krit}^2 \, v^2}{16 \alpha_{Dr}^4 \, R^2}}$$

(2)

Re_{krit} is the critical Reynolds number $(\mathrm{Re} = 2300)$. v is the fluid kinematic viscosity, ρ is the fluid density and α_{Dr} is the flow coefficient [23] . To investigate the presented control scheme the knee mechanism is built up in a test rig (Figure 6).

Two additional pneumatic actuators are implemented into the rig. In this way, it is possible to apply forces to simulate the ground reaction force and so, to change the knee angle while simultaneously control the desired movement of the gait cycle. Displacement sensors inside of the cylinders of the knee mechanism measure the piston positions that allow the calculation of the knee angle. The test rig is controlled with a dSpace rapid prototyping system. Sample results of a kinematic gait cycle simulation are shown in Figure 7. The figure shows the piston movement, as well as the desired and the measured knee angle in a gait cycle. The results generated by the test rig show that a gait cycle can be performed with the developed control scheme. The piston movement occurs in counter direction as expected.

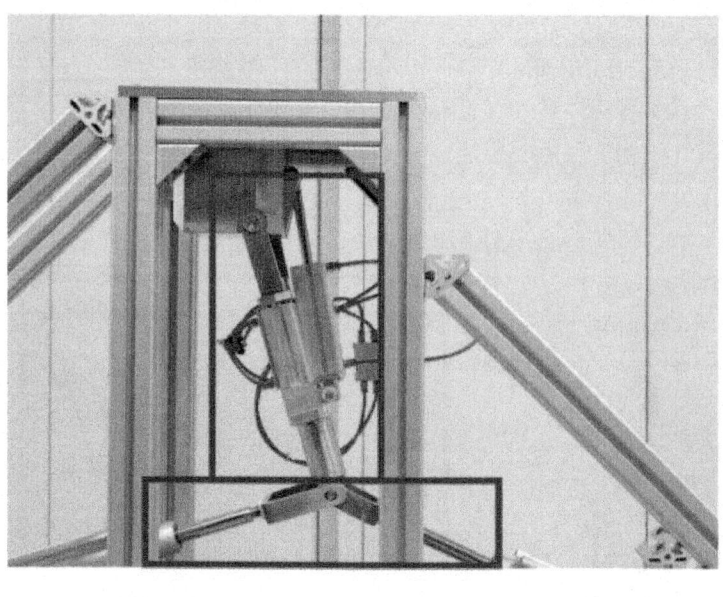

(a)

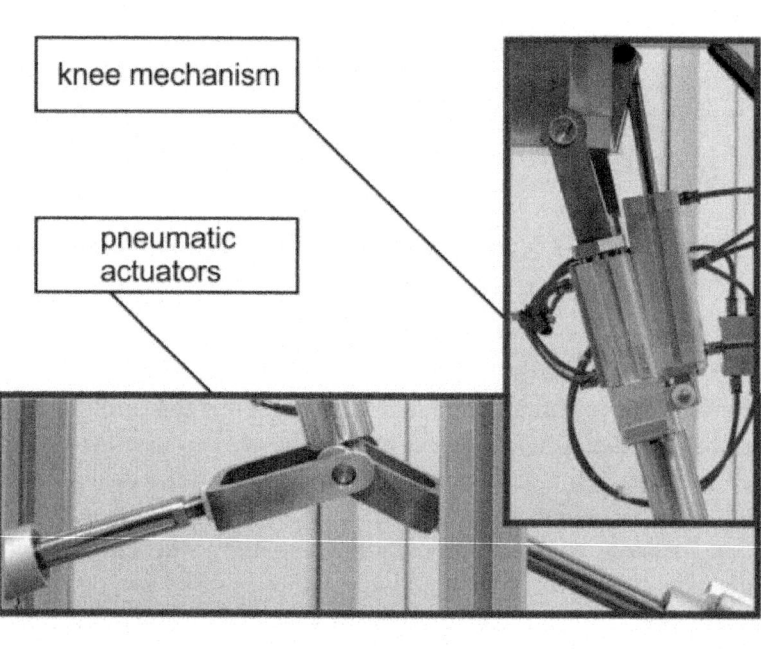

(b)

Figure 6: Test bed to investigate the knee mechanism.

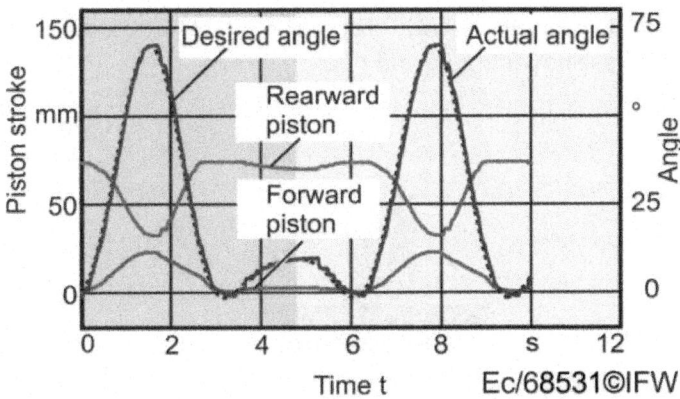

Figure 7: Results of the verifications with the test rig.

COMPENSATION OF LATERAL FORCES

Influence of Lateral Forces

The patient's center of gravity is not directly above the knee joint. Thus, the displacement between the patient's center of gravity and the knee d causes torque (T_{res}) acting on the suspension fork of the knee mechanism (see Figure 8). This torque is amplified by long lever arms, e.g. if the patient stumbles.

The difference of the lateral forces F_2, and F_3, resulting from the torque (T_{res}) leads to different movements of the pistons in the cylinders 2 and 3 and might jam those. Thus, the lateral forces must be absorbed, either mechanically or hydraulically by generating hydraulic forces which counteract the lateral forces and that way eliminate the torque T_{res}. The presented knee mechanism is designed to counteract the lateral forces using a linear bearing inside the hydraulic cylinders, as shown in Figure 8. The linear bearing has a height of 10 mm, an inner diameter of 13 mm and the outer diameter is 17 mm. This bearing is very light and space-saving, but, in the case of very high lateral forces, it might be too small to counteract all forces. A taller and more robust bearing would remedy this problem, but would also increase the necessary space and weight. In particular additional components outside the cylinders would have a significant stake of the prosthesis weight. Further weight reduction is attainable by combining the mechanical support structure with a new concept to reduce lateral forces hydraulically. In that case, additionally to the mechanical bearings, the hydraulic valves are controlled to generate hydraulic forces to compensate the lateral forces. The lateral forces to be compensated, are calculated as shown in (3), and (4):

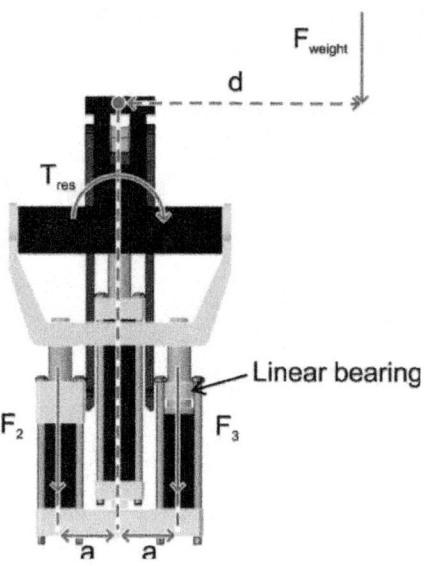

Figure 8: Force influence due to the patients weighty.

$$T_{res} = F_{weight} d = -F_2 a + F_3 a \tag{3}$$

$$F_{weight} = F_2 + F_3 \tag{4}$$

However, since valves are passive elements, only damping forces (forces against the moving direction of the pistons) can be generated. Generating a force in moving direction, however, would require additional actuators. The necessary direction of the compensation force depends on the current load case, as shown in (5), and (6):

$$F_{weight}\left(\frac{d}{a}-1\right) = -2F_2 \tag{5}$$

$$F_{weight}\left(\frac{d}{a}+1\right) = 2F_3 \tag{6}$$

If the distance between the line of action of the weight force d is less than the distance between the pivot of the folk and the hydraulic cylinder a, then both compensation forces (F_2, and F_3) would act against the moving direction of the pistons and the whole torque T_{res} can be eliminated by controlling the valves. If $d > a$, however, the compensation force F_2 would have to act in the same direction as the piston movement. Since a force in that direction cannot be generated in a passive system, F_2 is set to 0 and a combination of mechanical and hydraulic absorption of lateral forces is required (7).

$$T_{res} = F_{weight} d = F_3 a + T_{mech} \tag{7}$$

T_{mech} is the torque, generated by the linear bearings. To estimate the actual lever arm d, the standard ISO 10328 [10] is used. The different test cases expect a lever arm d between 35 mm and 57 mm. The distance between the pivot of the fork and the cylinder a is 34 mm. Thus, in the worst test case (d = 57 mm), only 60% of the torque can be absorbed hydraulically and the linear bearing has to be designed to withstand 40% of the torque T_{res}.

Control Based Absorption of Lateral Forces

In the following, two approaches for a control based compensation of the lateral forces are presented. The first approach is an open loop control scheme (feed forward), wherein the forces are detected by force sensors and the necessary cross sections of the valves are calculated by a model of the hydraulic mechanical couplings (Figure 9(a)). The second control scheme is a closed loop control (feed back), wherein the difference of the piston position is detected and corrected by the valves (Figure 9(b)).

To realize the feed forward approach, the mathematical derivation of the mechanical and the hydraulic system in Figure 2 is presented while the mechanical system is loaded with the three forces F_1, F_2, and F_3.

The equations of motion of the cylinders are

$$m_1 \ddot{x}_1 = P_{12} A_2 - p_1 A_1 - F_1 \tag{8}$$

$$m_2 \ddot{x}_2 = P_{22} A_2 - p_1 A_1 - F_2 \tag{9}$$

$$m_2 \ddot{x}_3 = P_{32} A_2 - p_1 A_1 - F_3 \tag{10}$$

where m_1 is the mass of the piston in cylinder 1 and m_2 is the mass of the pistons in cylinders 2 and 3. P_1, P_{12}, P_{22}, P_{32} are the pressures in the cylinder chambers (Figure 2(c)). A_1 and A_2 are the piston surfaces. The forces F_1, F_2, and F_3 are the loads acting on the piston rods Figure 9(b)). Due to the incompressibility of the fluid, stiffness effects can be neglected. The volume flow $\dot{V}_1$ can be calculated as follows

$$\dot{V_1} = -\dot{x}_1 A_1 = \dot{x}_2 A_1 + \dot{x}_3 A_1 \tag{11}$$

Therefore, the relation between position velocities is

$$-\dot{x}_1 = \dot{x}_2 + \dot{x}_3 \tag{12}$$

The damping effect is provided by the valves 1 and 2. The mathematical description of the valves is analogous to a hydraulic resistor. Macia and Thaler describe a hydraulic resistor R as the ratio of pressure difference Δp and volume flow $\dot{V}$ [24] .

$$R = \frac{\Delta p}{\dot{V}} \tag{13}$$

Using Equations (8)-(13) the dynamic behavior of the undamped hydraulic mechanical system can be described in the following way

$$\ddot{x}_1 \left(2m_1 + m_2\right) = A_2^2 \left(R_1 \dot{x}_2 + R_2 \dot{x}_3\right) - 2F_1 + F_2 + F_3 \tag{14}$$

Based on that general system equation, the necessary relation between the resistors R_1 and R_2 of the valves to compensate the lateral forces is derived as follows.

The pressure balance is calculated by (9) and (10):

$$m_2 \left(\ddot{x}_3 - \ddot{x}_2\right) = A_2 \left(p_{32} - p_{22}\right) + F_2 - F_3 \tag{15}$$

The pressure balance

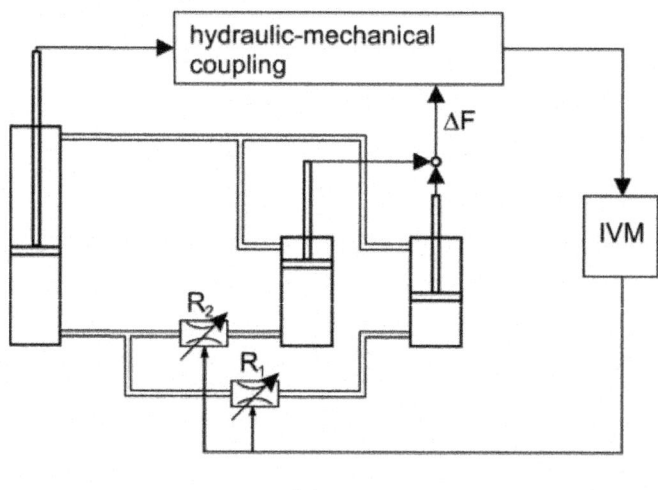

(a)

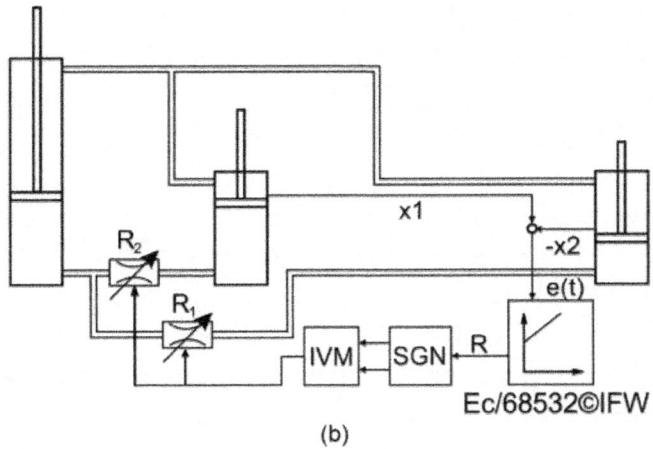

(b)

Figure 9: Control based approach to compensate lateral forces. (a) Feed forward control; (b) Feed back control.

$$p_{32} - p_{22} = R_1 \dot{V}_1 - R_2 \dot{V}_2 \tag{16}$$

As well as (13), and (15) lead to the dynamic behavior of the description of the dynamic coupling between the pistons.

$$\ddot{x}_3 - \ddot{x}_2 = \frac{1}{m_2} \left(A_2^2 \left(R_1 \dot{x}_2 - R_2 \dot{x}_3 \right) + F_2 - F_3 \right) \tag{17}$$

The pistons are constrained to equal speed:

$$x_2 = x_3, \quad \dot{x}_2 = \dot{x}_3 \quad \text{and} \quad \ddot{x}_2 = \ddot{x}_3 \tag{18}$$

The requirement (18) as well as (12) and (17) gives the relation between R_1 and R_2, which enables the compensation of lateral forces

$$R_2 - R_1 = 2 \frac{F_3 - F_2}{A_2^2 \dot{x}_1} \tag{19}$$

As an alternative to the feed forward approach, a feed back approach is presented, which determines the necessary cross sections of the valves to compensate lateral forces from the difference of the piston positions of cylinders 2 and 3 by a PI control element. Therefore, the positions of the piston rods of the cylinders are measured by displacement sensors and the difference of the two positions $e(t)$ is equal to the error to be compensated. Based on this error, the PI controller calculates the correcting variable R to compensate the lateral forces. The demanded cross section of the valves A_v is calculated from the correcting variable R by the inverse model of the valves (IVM).

For the investigation of both approaches a Matlab/Simulink model of the knee mechanism has been developed. The mechanical part is implemented by the SimMechanics toolbox, the hydraulic part is realized by the SimHydraulics toolbox. In the simulation model the three pistons of the knee mechanism are loaded by different force steps. All force step occurs at the same time (after 0.07 s).

F_1 is loaded with 1000 N, F_2 is loaded 800 N, and F_3 is loaded with 200 N. These test signals are chosen, because they show a high difference between the lateral forces F_2 and F_3. As a result, the three pistons move differently. The mechanical coupling of the pistons of cylinder 2 and 3 via the suspension fork as well as the mechanical support structure is not considered in these investigations.

To be able to compare and evaluate the approaches to compensate lateral forces, an assessment scheme is developed. The comparison values are the error surface and the maximum error (Figure 10). The goal is to minimize both values. Figure 11 shows the results. Analyzed are the feed forward approach (FF), the feedback approach (FB) with P and PI control element as well as combinations of both.

The column chart in Figure 11(a) a shows that the feed forward approach (FF) gives very accurate results.

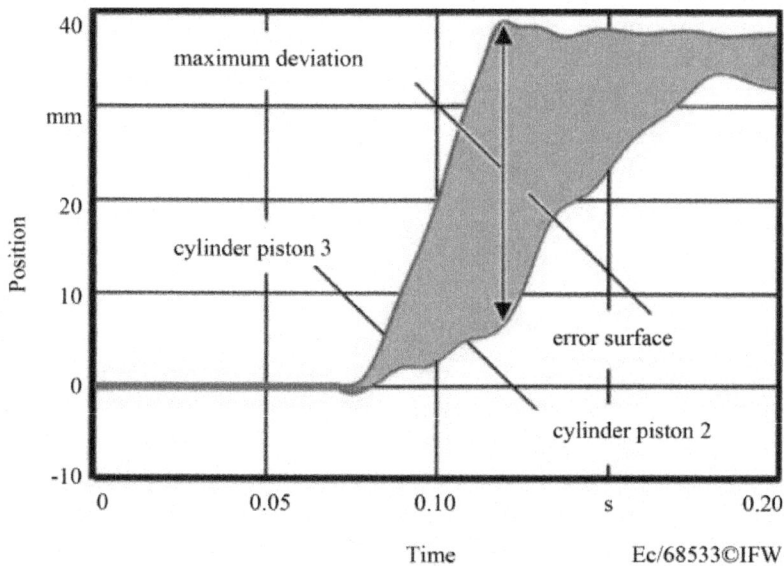

Figure 10: Assessment scheme.

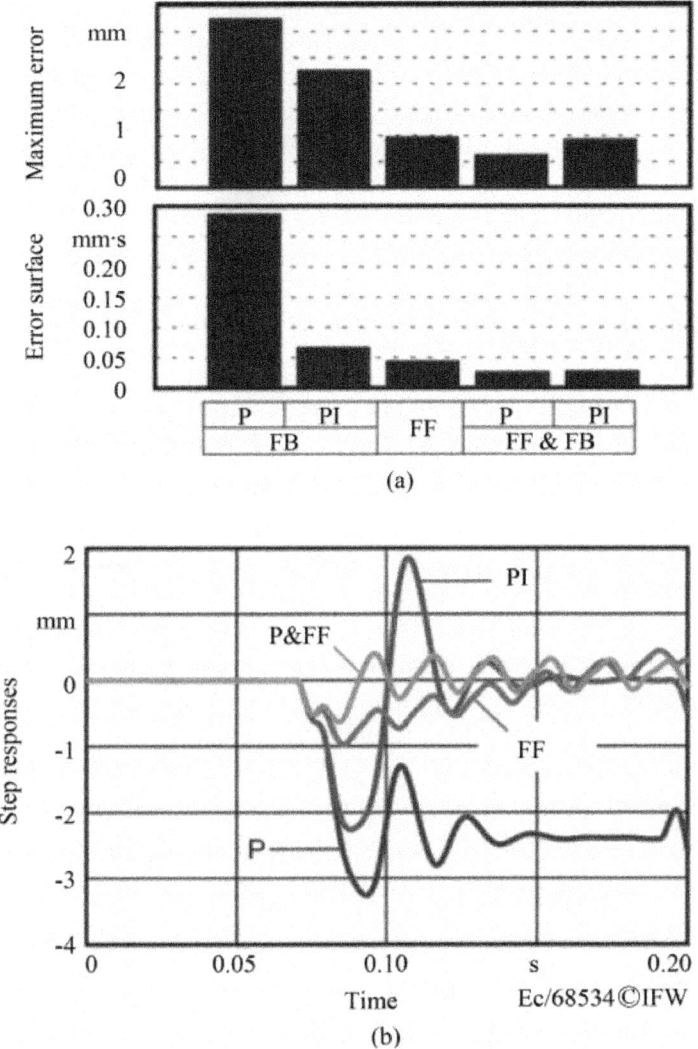

Figure 11: Results of the compensation of lateral forces.

The feed back approach (FB) based on a P-control element is less accurate. The closed loop approach based on a PI-controller starts as accurately as the P-control element which leads to almost the same maximum error (Figure 11(b)). The I-control element, however, eliminates the steady state error which leads to an error surface less than the P-control element. The best results are achieved by superposition of the open loop control and the closed loop control.

The open loop control approach shows better results for two reasons: On the one hand, the closed loop control approach because a model of the process is implemented. On the other hand, that approach includes additional process information given by the force sensors. Thus, the control is able to act before an error occurs. The closed loop control approach, however, reacts after an error has occurred which leads to higher error values. The presented investigations are based on ideal assumptions. To evaluate the actual system behavior, it is necessary to consider delays of the valves, model inaccuracies, measurement inaccuracies and inaccuracies in the valves adjustments. These evaluations are presented in Section 3.3.

Investigation of Dynamic Influences and Interferences

The analyses in Section 3.2 assume ideal conditions. The actual valves, however, do not act immediately but delayed. This delay is approximated by a first order low pass characteristic. Figure 12 shows the maximum deviation in response to the time constant of the valves.

For short time constants the open loop control approach shows the best results. In this case, the process model is very accurate. For higher time constants the error increases rapidly, because the model of the process does not consider the valve behavior. However, for higher time constants, the closed loop approach shows better results due to the subsequent elimination of the additional error, caused by the dynamic behavior of the valves. The best results are achieved by superposition of both approaches. Current high response valves achieve a switching frequency of 125 Hz and a switching time of less than 3 ms, which is enough to generate good results as shows in Figure 12.

In summary it has been shown that the hydraulic concept is able to absorb lateral forces, if the pistons of cylinder 2, and cylinder 3 are loaded with forces with a positive sign ($F_2 > 0$, $F_3 > 0$). In combination with the mechanical bearings, all loads, based on [10] can be eliminated.

ACTIVE LEG LENGTH VARIATION

Mathematical Description

The active variation of the leg length offers advantages for the patient. After an amputation the amputee is not able to lift his foot anymore. With the active variation of the leg length the patient is able to obtain more space between the moving foot and the ground. Thus, risks of collision and stumble can be reduced. The necessary energy to change the leg length is generated by the gait cycle. Thus, no external source of energy is required.

As depicted in Section 2 the movement of the pistons inside of the forward and the rearward cylinders in opposite direction changes the knee angle.

Uniform movement of all three pistons, however, leads to the variation of the leg length. To enable uniform movement, the hydraulic scheme, depicted inFigure 2(c) needs to be expanded by an additional volume chamber to compensate the volume of the piston rods. The expanded hydraulic scheme is shown in Figure 13. In order to simplify the following calculations the valves 1 and 2 are summarized to the valve b. In that case the pressures in the chambers as well as the piston rods in the forward cylinders are equal.

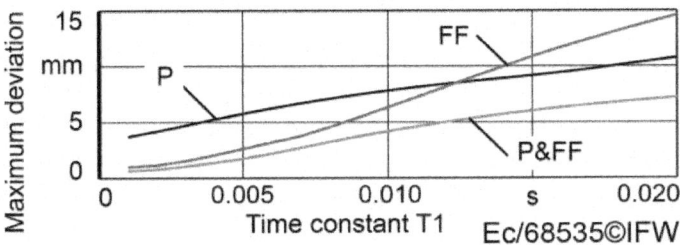

Figure 12: Influence of the valves dynamic.

This hydraulic scheme enables piston movements in opposite direction to change the knee angle as well as uniform motion to change the leg length. Superimposed motion of the angle and leg length is also possible. The directions of the movement of the pistons are adjusted by the two valves a and b by adjusting the cross section of the valves.

The shortening of the leg length happens in the stance phase and is driven by the weight of the patient. All three pistons move in negative direction $(-x_1, -x_2)$ and the hydraulic fluid fills the chamber (compensation element). Additionally, the weight of the patient compresses a spring energy storage. The stored energy is used to extend the leg length in the end of the swing phase.

The valve b damps the opposite movement of the pistons of the backward and the forward cylinders and that way it damps the knee angle, while the valve a damps the uniform motion of the cylinders and controls the leg length variation. The mathematical description of the expanded hydraulic scheme is shown as follows:

$$\dot{V}_a = A_k \cdot \dot{x}_k \tag{20}$$

$$\dot{V}_b = A_2 \cdot \dot{x}_2 \tag{21}$$

$$R_b = \frac{p_1 - p_2}{\dot{V}_b} \tag{22}$$

$$R_a = \frac{p_1 - p_k}{\dot{V}_a}$$

(23)

where R_a and R_b are the hydraulic resistors of the valves a and b. The necessary value of R_b depends on the demanded motion of $\dot{x}_2$ and is calculated by the Equations (21)-(22)

$$R_b = \frac{p_1 - p_2}{A_2 \dot{x}_2}$$

(24)

The necessary value of R_a is calculated by the Equation (20), and Equation (23) as follows:

$$R_a = \frac{p_1 - p_2}{A_k \cdot \dot{x}_k}$$

(25)

The Equations (24)-(25) enable the calculation of the values of the hydraulic resistances R_a, and R_b. These values lead to the demanded damping of the knee angle and the demanded leg length variation at the same time. The relation between hydraulic resistor and the adjustable cross section of the valve as well as the relation between positions and the knee angle is depicted in Section 2.2.

Simulation of a Gait Cycle with Active Leg Length Variation

The scheme, presented in Figure 13 is investigated with a simulation model. The control scheme as well as the inverse and direct kinematics are realized by Matlab/Simulink. This model allows simulating a whole gait cycle as depicted in Figure 14.

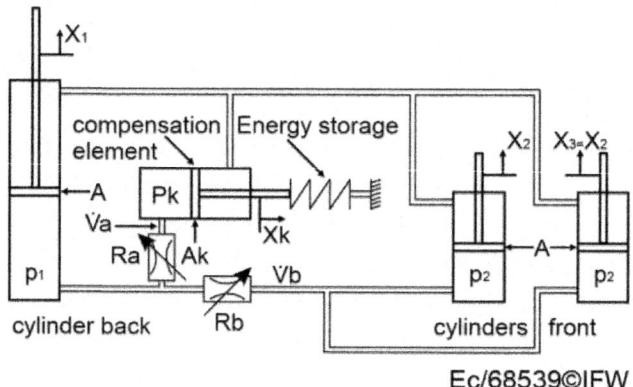

Ec/68539©IFW

Figure 13: Extended hydraulic scheme.

The actual ground reaction force and the influence of the torque from the hip are considered in this simulation model. The simulation shows that a gait cycle with active leg length variation can be realized with the presented knee mechanism. Figure 14 shows a gait cycle with leg length variation. The leg is shortened within the stance phase using the patient's weight and it is lengthened within the swing phase using the force of the spring.

In the presented case, the leg length is varied by 10 mm. However, the curve of the leg length is just an example to show, that the actual leg length follows the desired leg length very well. The necessary curve of leg length variation is an anatomic challenge which has not been investigated yet.

Alternative to the presented hydraulic scheme, it is possible to realize leg length variation without an additional energy store. In that case, within the stance phase, only the piston inside the backward cylinder is moving and that way the knee angle as well as the leg length is changing, while a volume chamber is filled. To extend the leg length within the swing phase, the kinetic energy of the leg is used.

Investigation of Dynamic Influences

The presented simulation assumes ideal conditions. Additionally, the influence of the low pass behavior of the valves is investigated using the assessment scheme, presented in Section 3.2. Here, the influence of the accuracy of the leg length variation of 10 mm is explored. The knee angle is not considered because the leg length variation is more sensitive than the knee angle damping.

Figure 15 shows the influence of the low pass behavior. The error of the leg length increases approximately linearly. A time constant of 0.02 seconds leads to an error of the leg length of 1 mm or 10% of the desired variation.

In general the influence of the valve behavior of the gait cycle as well as of the leg length variation is less than of the compensation of the lateral forces, because the lateral forces appear immediately.

CONCLUSIONS

This paper provides an overview into the development of a novel mechanism for knee disarticulation prostheses. Common approaches of artificial knee joints consist of a mechanical support structure to transfer forces and an additional hydraulic damping element. In contrast, for this approach, no mechanical support structure is necessary. Instead, the hydraulic system is designed to transfer the weight force, as well as to reduce the lateral forces. This makes the mechanism more flexible; it helps to reduce weight and allows additional applications, like the active variation of the leg length within a gait cycle. First, a method to compensate lateral forces hydraulically is presented. Here the control of the joint mechanism is augmented with a mathematical model for the description of the mechanical hydraulic coupling. Via a controlled operation of the valves the lateral forces are absorbed by the hydraulic system. Furthermore, a control scheme is presented which enables the active variation of the leg length with this novel knee mechanism. The patient's weight force is used to shorten the leg length in the stance phase. Hereby a spring is compressed. Using the spring energy, a leg elongation in the swing phase is achieved.

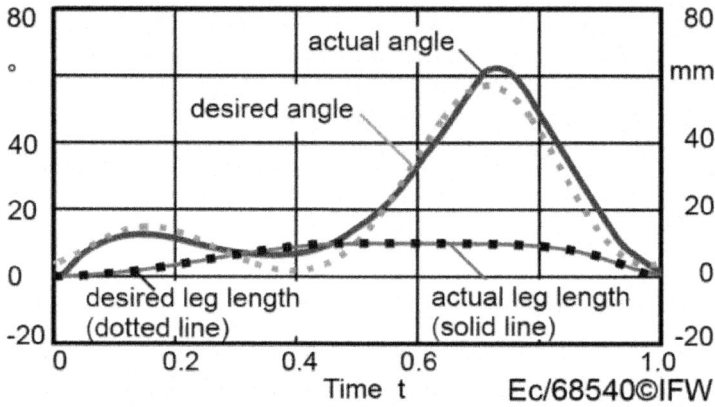

Figure 14: Simulation of a gait cycle with leg length variation.

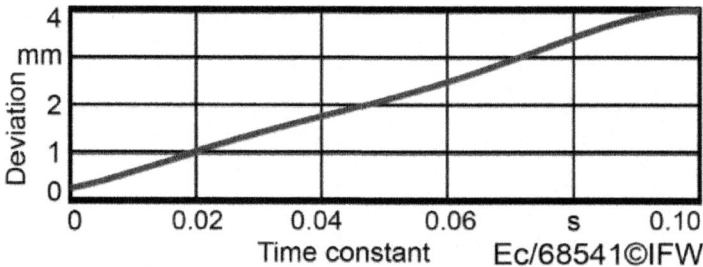

Figure 15: Influence of the valve dynamic of the leg length variation.

Since the only source of energy is the weight force, a leg length variation of 20 mm can be realized without additional actuators. By using the kinetic energy in the swing phase, it is also possible to realize active leg length variation without an additional spring energy store.

With the new prosthesis, a knee angle of 136° is attainable and it is designed to withstand a patient's weight of 125 kg. The prosthesis complies with the standard ISO10328 and weighs only 2.5 kg.

ACKNOWLEDGEMENTS

The research studies, presented in this paper, have been part of the cooperation project MultiPro. The authors would like to thank the Federal Ministry for Research and Education (BMBF) for funding this project.

REFERENCES

1. Hagberg, E., Berlin, Ö.K. and Renström, P. (1992) Function after Through-Knee Compared with Below-Knee and Above-Knee Amputation. Prosthetics and Orthotics International, 16, 168-173.

2. Behr, J., Friedly, J., Molton, I., Morgenrot, D., Jensen, M.P. and Smith, D.G. (2009) Pain and Pain-Related Interference in Adults with Lower-Limb Amputation: Comparison of Knee-Disarticulation, Transtibial and Transfemoral Surgical Sites. Journal of Rehabilitation & Development, 46, 963-972. http://dx.doi.org/10.1682/JRRD.2008.07.0085

3. Greitemann, B. (2005) Kniegelenknahe Amputationen. In: Wirth, J.C. and Zichner, L., Eds., Orthopädie und Orth- opädische Chirurgie, Georg Thieme Verlag, Stuttgart, 481-491.

4. Baumgartner, R.F. (1979) Knee Disarticulation versus Above-Knee Amputation. Prosthetics and Orthotics International, 3, 15-19.

5. Mensch, G. (1983) Physiotherapy Following Through-Knee Amputation. Prosthetics and Orthotics International, 7, 79-87.

6. Stark, G. (2004) Overview of Knee Disarticulation. American Academy of Orthotists & Prosthetists, 16, 130-137.

7. O'Connor, R.S. (1999) Prosthesis for Long Femur and Knee Disarticulation Amputation. Patent 5895430, USA.

8. Kramer, S., Srinivasan, S. and Swanson, V. (1998) Knee Joint Mechanism for Knee Disarticulation Prosthesis. Patent 5746774, USA.

9. Denkena, B., Simon, S. and Zager, H. (2011) Knee Exarticulation Prosthesis for Use as Hydraulic Knee Joint Prosthesis for Human Leg, Has Hydraulic Assemblies Causing Motion of Artificial Replacement Member of Leg Sections around Rotary Axis by Movement in Opponent Arrangement. Patent DE102009056074A1, Germany.

10. ISO10328:2006 (2006) Prosthetics-Structural Testing of Lower-Limb Prostheses-Requirements and Test Methods.

11. Perry, J. and Burnfield, J. (1992) Gait Analysis: Normal and Pathological Function. Slack, New York.

12. Whittle, M.W. (1996) Gait Analysis: An Introduction. Butterworth-Heinemann Ltd., Oxford.

13. Al-Hamadi, A., Andres, S., Berndt, D., Calow, R., Lehmann, C., Michaelis, B., Schünemann, S. and Urbansky, U. (2004) 3D-Bewegungsanalyse in der "Neuro-Medizintechnik". Ph.D. Dissertation, Otto-von-Guericke-Universität Magdeburg, Magdeburg.

14. Götze, C., Sippel, C., Rosenbaum, D., Heckenberg, L. and Steinbeck, J. (2003) Objective Measures of Gait Following Revision Hip Arthroplasty. First Medium-Term Results 2.6 Years after Surgery. Zeitschrift für Orthopädie und ihre Grenzgebiete, 141, 201-208.

15. Seichert, N., Erhart, P. and Senn, E. (1997) Die Etablierung der instrumentierten Ganganalyse (IGA) als Verfahren zur unmittelbaren klinikrelevanten Gangbeurteilung. Darstellung der propulsiven und bremsenden Muskelaktivitäten beim Gehen. Physikalische Medizin, Rehabilitationsmedizin, 7, 1-11. http://dx.doi.org/10.1055/s-2008-1061850

16. Kramers-de Quervain, I.A., Stüssi, E. and Stacoff, A. (2008) Ganganalyse beim Gehen und Laufen. Schweizerische Zeitschrift für Sportmedizin und Sporttraumatologie, 56, 35-42.

17. Leuchte, S. and Luchs, A. (2006) How Symmetrical or Asymmetrical Is the Physiological Gait in Respect of Age? Journal of Physical and Rehabilitation Medicine, 16, 96-102.

18. Hölscher, C. (2009) Auswirkungen Von Hohen Tibialen Umstellungs-Osteotomien auf den Gang und die Kniegelenks- Belastung. Ph.D. Dissertation, Medizinische Fakultät der Westfälischen Wilhelms-Universität Münster, Münster.

19. Fletcher, R. (1987) Practical Methods of Optimization. John Wiley & Sons Limited, West Sussex.

20. Gill, P.E., Murray, W. and Wright, M. (1982) Practical Optimization. Emerald Group Publishing Limited, London.

21. Sneddon, I.N. (1995) Fourier Transforms. Dover Publications, Inc., New York.

22. Clarke, J.A. (1985) Energy Simulation in Building Design. Butterwort-Heinemann, Oxford.

23. Crane Co., NY. (1982) Flow of Fluids through Valves, Fittings and Pipes. Metric Edition-SI Units, Technical Paper No. 410M, Crane Co., New York.

24. Macia, N.F. and Thaler, G.J. (2005) Fluid Systems-Hydraulic. In: Macia, N.F. and Thaler, G.J., Eds., Modeling & Control of Dynamic Systems, Thomson Delmar Learning, Clifton Park, 91-95.

CITATION

Chapter 8

JOULE HEATING AND THERMAL RADIATION EFFECTS ON MHD BOUNDARY LAYER FLOW OF A NANOFLUID OVER AN EXPONENTIALLY STRETCHING SHEET IN A POROUS MEDIUM

Jakkula Anand Rao[1], Gandamalla Vasumathi[1], and Jakkula Mounica[2]

[1]Department of Mathematics, Osmania University, Hyderabad, Telangana

[2]Department of Mathematics, NGRI, Hyderabad, Telangana

ABSTRACT

A numerical study on boundary layer flow behaviour, heat and mass transfer characteristics of a nanofluid over an exponentially stretching sheet in a porous medium is presented in this paper. The sheet is assumed to be permeable. The governing partial differential equations are transformed into coupled nonlinear ordinary differential equations by using suitable similarity transformations. The transformed equations are then solved numerically using the well known explicit finite difference scheme known as the Keller Box method. A detailed parametric study is performed to access the influence of the physical parameters on longitudinal velocity, temperature and nanoparticle volume fraction profiles as well as the local skin-friction coefficient, local Nusselt number and the local Sherwood number and then, the results are presented in both graphical and tabular forms.

INTRODUCTION

The industrial processes like hot rolling, wire drawing, spinning of filaments, metal extrusion, crystal growing, glass fibre production, paper production, cooling of a large metallic plate in a bath, which may be an electrolyte, etc. to require the study of flow and heat transfer over a stretching surface. In all these cases, the quality of final product depends on the surface heat transfer rate and the skin friction coefficient. So this study has gained considerable attention in the recent years. Choi [1] was the first to sort heat transfer enhancement upon the invention of "nanofluid". These fluids are engineered colloidal suspensions composed of nanoparticles in a base fluid. Eastman [2] observed thermal conductivity enhancement in nanofluids. Metals (Al, Cu), oxides (Al_2O_3, TiO_2 and CuO), carbides (SiC), nitrides (AlN, SiN), or nonmetals (graphite, carbon nanotubes) and conductive fluids, such as water or ethylene glycol, or oil, other lubricants, bio-fluids, polymer solutions as base fluids are used in the manufacturing of nanofluids. 5% volume fraction of nanoparticles in these fluids ensure effective heat transfer enhancements which help them to exhibit enhanced thermal conductivity and the convective heat transfer coefficient compared with the base fluid. Routbort [3] found that the typical thermal conductivity enhancements and heat transfer coefficient enhancements are in the range of 15% - 40% and up to 40% over the base fluids. But still there must be other mechanisms to attribute to higher thermal conductivity. Increase in thermal conductivity cannot be considered as the sole reason. Here a survey conducted by Buongiorno [4] on convective transport in nanofluids implied that energy transfer by dispersion of nanoparticles was negligible. His model could not explain the abnormal heat transfer coefficients. The boundary layer has different properties due to the effect of temperature and thermophoresis. There may be a decrease in viscosity in the boundary layer, which will lead to heat transfer enhancement. Many other literatures support Buongiorno's explanation. An excellent assessment of nanofluid physics and developments had been provided by Cheng [5] and Ali [6] . Buongiorno and Hu [7] observed that although convective heat transfer enhancement had been suggested to be due to the dispersion of the suspended nanoparticles, this effect was too small to explain the observed enhancement.

It is often assumed in the problems of boundary layer flow over a stretching surface that the velocity of the stretching surface is linearly proportional to distance from the fixed origin. However, Gupta [8] had argued against the linearity condition. In the real world problem, the stretching of plastic sheet may not necessarily be linear. For example an exponentially stretching sheet was considered by Sanjayanand [9] . Heat and mass transfer on its boundary layers are investigated. Suction and heat transfer characteristics were addressed by Youn [10] . Heat transfer in a viscoelastic boundary layer flow over a stretching sheet with viscous dissipation and non-uniform heat source was studied by Subhas [11] . Thermal radiation effect on the flow was examined by Sohail Nadeem [12] . Anuar Ishak [13] studied the MHD boundary layer flow due to an exponentially stretching surface having the effect of radiation. Effect of magnetic field on boundary layer flow and heat transfer of a dusty fluid with an exponential temperature distribution on the exponentially stretching sheet was considered by Al-odat [14] .

Hitesh Kumar [15] worked on the heat transfer MHD boundary-layer flow through a porous medium. Gopi Chand [16] considered an unsteady stretching surface in a porous medium and explained the viscous dissipation and Radiation effects on MHD flow over it. Flow through a porous medium bounded by a vertical surface in presence of hall current was explained by Sudhakar [17] . joules DISSIPATION over a nonlinear vertical stretching porous sheet was introduced by Subhas Abel [18] . M.M. Hamza [19] provided the study of Oscillatory Flow through a porous medium. P.R. Sharma [20] worked on a porous stretching sheet.

This paper provides the solution to the problem of flow and heat transfer of a nanofluid over an exponentially stretching porous sheet by considering the effect of chemical reaction, joule heating and thermal radiation parameters along with the suction parameter by adopting the Keller Box method.

FORMULATION OF THE PROBLEM

The present problem is based on a steady two-dimensional incompressible viscous laminar flow of an electrically conducting nanofluid over a permeable exponentially stretching sheet in a porous medium as shown in Figure 1. The X-axis is taken along the stretching surface in the direction of motion and the Y-axis is perpendicular to it. The flow is confined to $y > 0$. Two equal and opposite forces are applied along the X-axis. Keeping the origin fixed, the sheet is then stretched with a velocity $U_w(x)$, varying exponentially with the distance from the slit. It is assumed that the surface temperature T_w and nanoparticle fraction C_w are constants and they differ from the ambient values of temperature and nanoparticle fraction T_∞ and C_∞, respectively. The usual boundary layer approximations are still applicable. The governing equations of continuity, momentum, energy and concentration in Cartesian coordinates x and y are:

$$\frac{\partial u}{\partial x} + \frac{\partial v}{\partial y} = 0$$

(1)

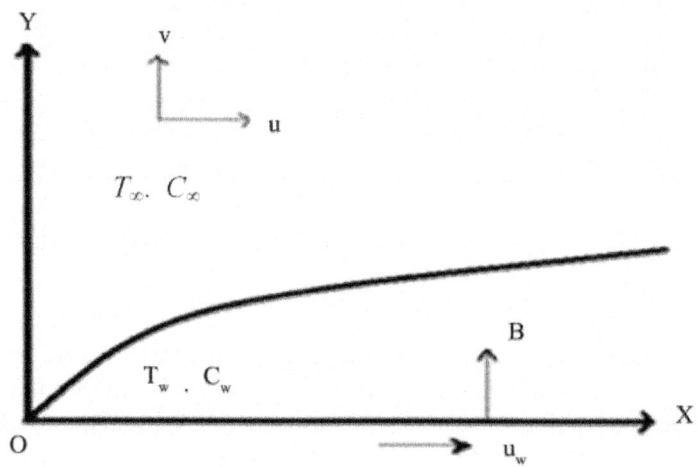

Figure 1: Physical modal and coordinate system.

$$u\frac{\partial u}{\partial x}+v\frac{\partial u}{\partial y}=\upsilon\frac{\partial^2 u}{\partial y^2}-\frac{\sigma B_0^2 u}{\rho}-\frac{\upsilon}{K_1}u$$

(2)

$$u\frac{\partial T}{\partial x}+v\frac{\partial T}{\partial y}=\alpha_m\frac{\partial^2 T}{\partial y^2}-\frac{1}{(\rho c)_f}\frac{\partial q_r}{\partial y}+\tau\left[D_B\frac{\partial C}{\partial y}\frac{\partial T}{\partial y}+\frac{D_T}{T_\infty}\left(\frac{\partial T}{\partial y}\right)^2\right]+\frac{\sigma B_0^2 u^2}{\rho c_p}$$

(3)

$$u\frac{\partial C}{\partial x}+v\frac{\partial C}{\partial y}=D_B\frac{\partial^2 C}{\partial y^2}+\frac{D_T}{T_\infty}\frac{\partial^2 T}{\partial y^2}-k_1(C-C_\infty)$$

(4)

where, u & v are the velocity components in X and Y-directions respectively, ρ-density of the nanofluid, B_0-in-duced magnetic field, σ-electrical conductivity, υ-kinematic viscosity, K_1-thermal conductivity, α_m-thermal diffusivity parameter, $(\rho c)_f$-the heat capacitance of the base fluid, τ-ratio between the effective heat capacity of the nanoparticle material and heat capacity of the nanofluid, D_B-the Brownian diffusion coefficient, D_T-the thermophoresis coefficient, q_r-radiative heat flux, k_1-chemical reaction parameter, c_p-specific heat capacity of the nanoparticle, γ-chemical reaction parameter, α-the chemical reaction coefficient, T-temperature of the nanofluid, C-concentration of the nanofluid, T_w and C_w-the temperature and concentration along the stretching sheet, T_∞ and C_∞-the ambient temperature and concentration.

The boundary conditions are:

$$u=U_w(x),v=-V_w(x),T=T_w,C=C_w,\text{ at }y=0$$
$$u=0,T\to T_\infty,C\to C_\infty\text{ as }y\to\infty$$

(5)

By using the Rosseland approximation for radiative heat flux is defined as:

$$q_r = -\frac{4\sigma^*}{3K^*}\frac{\partial T^4}{\partial y}$$

(6)

where K^* is the mean absorption coefficient and σ^* is the Stefan-Boltzmann constant. Since the temperature differences within the flow field are assumed to be small, and then we linearize and expand T^4 into the Taylor series about T_∞, which after neglecting higher order forms takes the form.

$$T^4 = 4T_\infty^3 T - 3T_\infty^4$$

(7)

To examine the flow, the following transformations are used:

$$u = U_0 e^{\frac{x}{L}} f'(\eta), \eta = \sqrt{\frac{U_0}{2\upsilon L}} e^{\frac{x}{2L}} y, v = -\sqrt{\frac{\upsilon U_0}{2L}} e^{\frac{x}{2L}} \left[f(\eta) + \eta f'(\eta) \right]$$

$$\theta(\eta) = \frac{T - T_\infty}{T_W - T_\infty}, \phi(\eta) = \frac{C - C_\infty}{C_W - C_\infty}$$

(8)

To determine the velocity, temperature distribution and rate of heat and mass transfer in the above boundary layer (5), we solve the equations related to the stretching sheet problem to obtain the following similarity equations using (8). In deriving these equations, the external electric field is assumed to be zero and the electric field due to polarization of charges in negligible.

$$f''' + ff'' - 2(f')^2 - (M + G)f' = 0$$

(9)

$$\left(1 + \frac{4}{3}R\right)\frac{1}{Pr}\theta'' + Nb\phi'\theta' + Nt\theta'^2 + Hf''^2 + f\theta' = 0$$

(10)

$$\phi'' + Lef\phi' + \frac{Nt}{Nb}\theta'' - \gamma\phi = 0$$

(11)

The transformed boundary conditions take the following forms:

$$f(0) = S, f'(0) = 1, f'(\infty) = 0, \theta(0) = 1, \theta(\infty) = 0, \phi(0) = 1, \phi(\infty) = 0$$

(12)

where,

$$M = \frac{\sigma B_0^2}{\rho c}, G = \frac{\upsilon}{cK}, c = \frac{U_0 e^{\frac{x}{L}}}{2L}, H = M \cdot Ec, Ec = \frac{U_0 e^{\frac{x}{L}}}{c_p}, Le = \frac{\upsilon}{D_B}, \gamma = \frac{\alpha}{C_w - C_\infty}$$

$$Nt = \frac{\tau D_T(T_w - T_\infty)}{\upsilon T_\infty}, Nb = \frac{\tau D_B(C_w - C_\infty)}{\upsilon}, R = \frac{4T^*T_\infty^3}{k^*k}, Pr = \frac{\upsilon}{\alpha_m}, S = \frac{V_w}{\sqrt{\upsilon c}}$$

(13)

For the type of boundary layer flow, the skin-friction coefficient, heat transfer coefficient and mass transfer coefficients are important physical parameters.

They defined as:

$$C_f = \frac{\tau_w}{\rho U_w^2/2} = \frac{\mu\left(\frac{\partial u}{\partial y}\right)_{y=0}}{\rho U_w^2/2}, \; Nu_x = \frac{Lq_w}{k(T_w - T_\infty)} = -\frac{L\left(\frac{\partial T}{\partial y}\right)_{y=0}}{(T_w - T_\infty)}$$

$$\text{and } Sh_x = \frac{Lq_m}{k(C_w - C_\infty)} = -\frac{L\left(\frac{\partial C}{\partial y}\right)_{y=0}}{C_w - C_\infty}$$

(14)

The dimensionless forms of these parameters are:

$$\sqrt{Re_x}\,Cf_x = \sqrt{2}f''(0), \; Nu_x = -\left(1 + \frac{4}{3}R\right)\sqrt{\frac{Re_x}{2}}\theta'(0), \text{ and } Sh_x = -\sqrt{\frac{Re_x}{2}}\phi'(0)$$

(15)

where the surface shear stress $\tau_w = \mu\left(\frac{\partial u}{\partial y}\right)_{y=0}$, the surface heat flux $q_w = k\left(\frac{\partial T}{\partial y} + \frac{\partial q_r}{\partial y}\right)_{y=0}$, the surface mass flux $q_m = k\left(\frac{\partial C}{\partial y}\right)_{y=0}$ and $Re_x = \frac{U_w L}{\upsilon}$ is the Reynolds number with μ and k being the dynamic viscosity and the thermal conductivity, respectively. The numerical values of $f''(0)$, $\theta'(0)$ and $\phi'(0)$ are proportional to the local skin-friction coefficient, local Nusselt number and local Sherwood number respectively and these are presented by Table 1 for the values of the physical parameters.

NUMERICAL SOLUTION

Equations (9)-(11) subjected to the boundary conditions (12) are solved numerically using implicit finite difference method that is known as Keller Box in combination with the Newton's techniques as described by Cebeci and Bradshaw [21] . This method is unconditionally stable and has second order accuracy.

Table 1: Comparison of results for $-\theta'(0)$ with previous published works.

Pr	N.G. Rudraswamy [22]	Present work
0.72	0.6180	0.6152
1.0	0.7097	0.7093
1.5	0.7862	0.7862

The principal steps in using the Keller Box method are:

1.   Reducing higher order ODEs (systems of ODES) in to systems of first order ODEs;

12. Writing the systems of first order ODEs into difference equations using central differencing scheme;

3.   Linearizing the difference equations using Newton's method and wring it in vector form;

4.   Solving the system of equations using block elimination method.

In order to solve the above differential equations numerically, we adopt Matlab software which is very efficient in using the well known Keller Box method. In accordance with the boundary layer analysis, the boundary condition (12) at $\eta = \infty$ is replaced by $\eta = 5$, and the step size $\Delta\eta = 0.05$ is used to obtain numerical solution with five decimal place accuracy as the criterion of convergence. Obtained coupled ordinary non-linear Equations (9)-(11) are solved by Keller Box method for boundary condition (12). Accuracy of this numericalmethod shown in Table 1 is being validated by direct comparison with the numerical results reported by N.G. Rudraswamy [22] , Meraj Mustafa et al. [23] . The numerical comparisons of, $-\theta'(0)$, $-f''(0)$ and $-\phi'(0)$ for the values of Pr, Le, S, γ, R, Nt, Nb, M, G and H are shown in Table 2.

RESULTS AND DISCUSSION

Referring from Figure 2 larger Prandtl number has a relatively lower thermal diffusivity. This is because Pris defined as the ration of kinematic viscosity to thermal diffusivity. Thus an increase in Pr reduces thermal diffusivity; it decreases the thermal boundary layer thickness. Consequently, increases rate of heat transfer, and thereby increases the variation in the thermal characteristics. As expected, the variation in the temperature is more pronounced for smaller values of Pr than its larger values.

Figures 3-5 depict the effects of suction parameter S on velocity, temperature and concentration profiles, respectively at the boundary for exponentially stretching sheet. It is observed that velocity decreases significantly with increasing suction parameter where as fluid velocity is found to increase with blowing. It is observed that, when the wall suction (S > 0) is considered, this causes a decrease in the boundary layer thickness and the velocity field is reduced. Opposite behaviour is noted for blowing S < 0. In Figure 4, it is seen that temperature decreases with increasing suction parameter where as it increases due to blowing. Temperature overshoot is noted for blowing (S < 0) this feature prevails up to certain heights and then the process slowed down and at a far distance from the wall temperature vanishes. In Figure 5, it is observed that concentration decreases with increasing suction parameter where as it increases due to blowing.

Figure 6 shows the effect of Le on the dimensionless concentration for fixed values of other parameters. It is observed that for larger values of Le suppress the concentration profile i.e. inhibit nanoparticle species diffusion, as observed. There will be a much greater reduction in the concentration boundary layer thickness.

Figures 7-10 show that an abnormal increase in the concentration φ is found for a weaker Brownian motion Nb in fact an over shoot in the concentration function occurs Nt increases gradually. The effect is seen in the case of temperature profile. An appreciable increase in temperature profile is found for increasing values of Nt. In this case, an increase in temperature is found with increasing values of Nb.

Table 2. Showing results of $-\theta'(0)$, $-f''(0)$ and $-\phi'(0)$ for the values of Pr, M, S, Le, R, γ, G, H, Nt, and Nb.

Pr	Le	S	γ	R	Nt	Nb	M	G	H	θ(0)	f(0)	φ'(0)
0.72	10	1	0.01	0.01	0.45	0.45	0.3	0.5	0.06	-0.5992	-2.1252	-10.3194
1										-0.6919	-2.1252	-10.3194
1.5										-0.7677	-2.1252	-10.2214
	10									-0.5992	-2.1252	-10.3194
	15									-0.5934	-2.1252	-15.3972
	20									-0.5903	-2.1252	-20.4433
		-0.2								-0.1744	-1.4741	-1.0406
		0.2								-0.3017	-1.6648	-3.5434
		0.4								-0.3494	-1.7701	-5.1324
			0.1							-0.5991	-2.1252	-10.3317
			1							-0.5986	-2.1252	-10.4507
				0.1						-0.5648	-2.1252	-10.3431
				0.2						-0.5302	-2.1252	-10.3674
				0.3						-0.4992	-2.1252	-10.3895
					0.4					-0.6102	-2.1252	-10.2551
					0.8					-0.531	-2.1252	-10.5394
					1.2					-0.4669	-2.1252	-10.6288
						0.4				-0.6182	-2.1252	-10.3511
						0.8				-0.4838	-2.1252	-10.1787
						1.2				-0.3829	-2.1252	-10.1445
							0.5			-0.5965	-2.1865	-10.3149
							1.5			-0.5856	-2.4636	-10.2956
							5			-0.5626	-3.2155	-10.2502
								0.5		-0.5992	-2.1252	-10.3194
								1.5		-0.5875	-2.4115	-10.2991
								5		-0.5636	-3.1783	-10.2522
									0.06	-0.5992	-2.1252	-10.3194
									0.9	-0.4896	-2.1252	-10.4189
									1	-0.4765	-2.1252	-10.4308

Figure 11 and Figure 12 exhibits the nature of velocity field for the variation of magnetic parameter M. With increasing M, velocity is found to decrease but the temperature increases. As the Lorentz force opposes the motion of the fluid, much fluid is not entering to the boundary layer. It enforces the thickening of thermal layer.

As shown in Figure 13, it is observed that the nanoparicle volume fraction decreases with increasing values of chemical reaction parameter whereas the velocity and temperature profiles are not significant with the chemical reaction parameter.

As depicted in Figure 14, it is noticed that an increase in R yields an increase in the nanofluid's temperature, which leads to decrease in the heat transfer rate. Thus, the radiation should be at its minimum in order to facilitate the cooling process. All these physical behaviour are due to the combined effects of the strength of the Brownian motion and thermophoresis particle deposition.

Figure 15 and Figure 16, show effect of porosity parameter G on the temperature and velocity profiles, respectively. It is observed that the presence of the porous medium. The temperature profile whereas is reduces the velocity profile. This is because the porous medium inhibits the fluid not to move freely through the boundary layer. This leads the flow to increase thermal boundary layer thickness.

As shown in Figure 17 and Figure 18, it is observed that an increase in joule heating parameter H the profiles for temperature and concentration are decreasing.

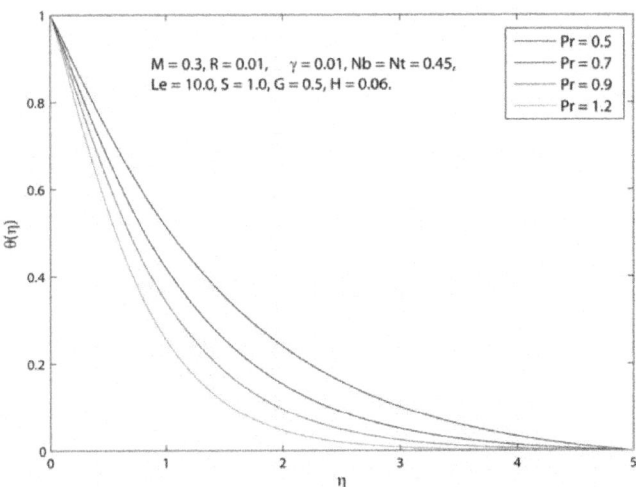

Figure 2: Variation of temperature $\theta(\eta)$ with η for several values of Prandtl number Pr.

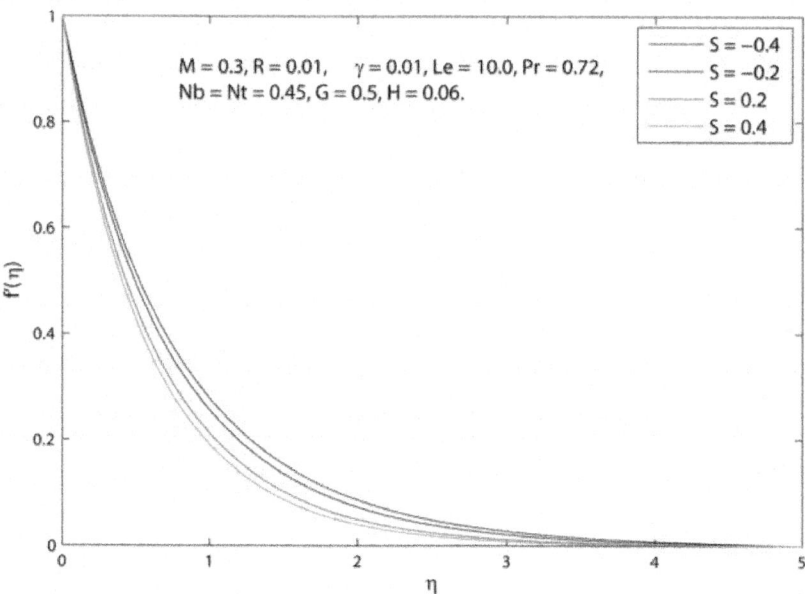

Figure 3: Variation of velocity $f'(\eta)$ with η for several values of suction parameter S.

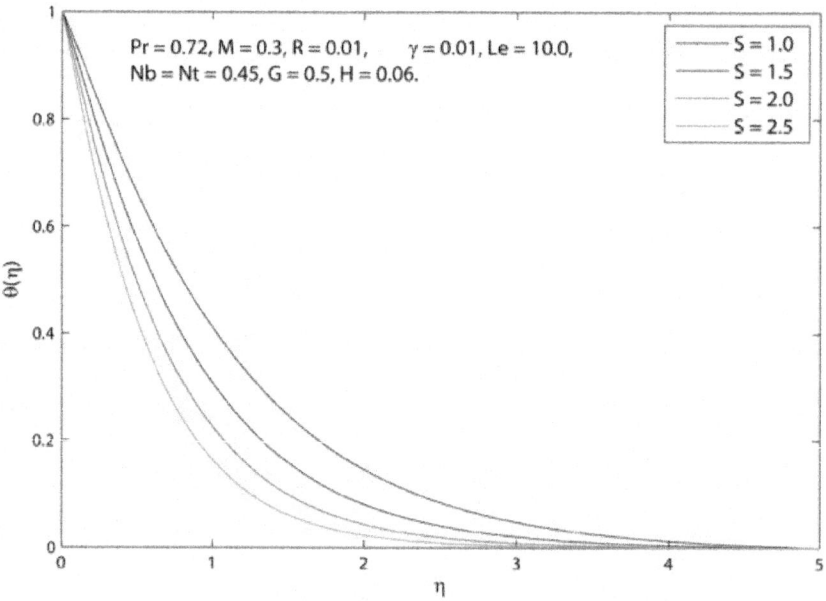

Figure 4: Variation of θ(η) with η for several values of suction parameter S.

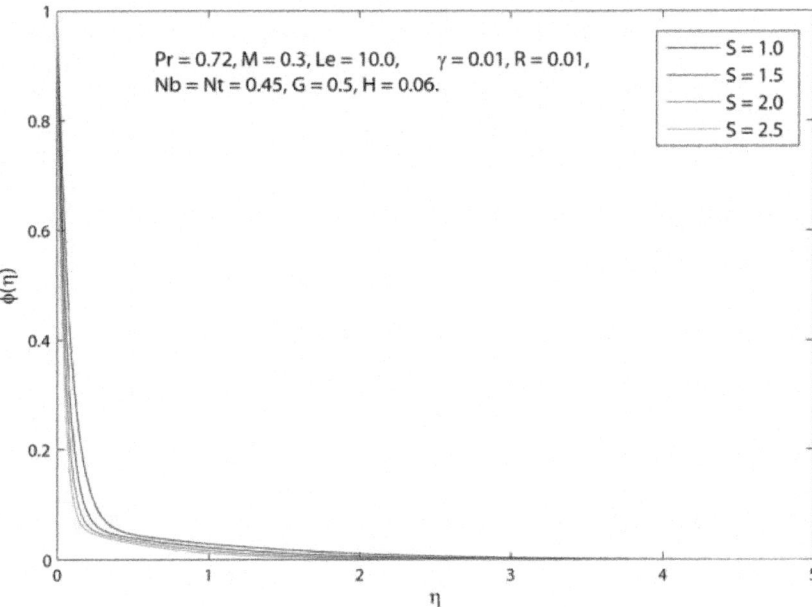

Figure 5: Variation of concentration φ(η) with η for several values of suction of S.

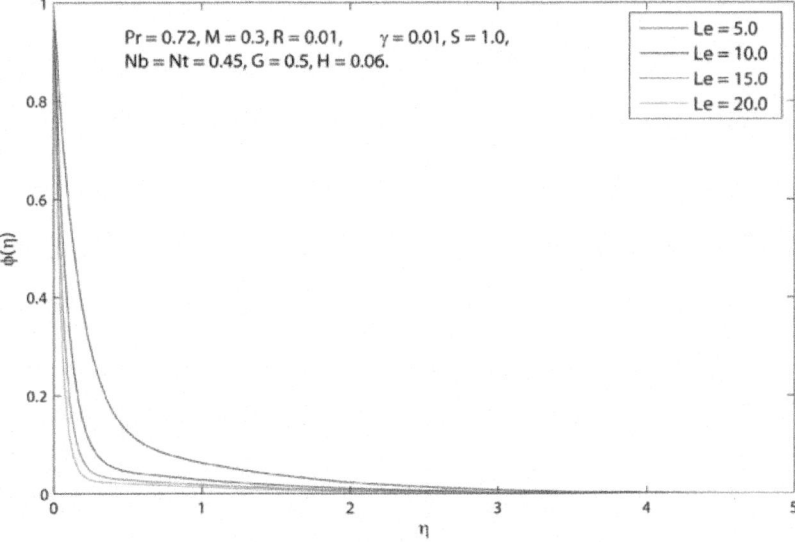

Figure 6: Variation of concentration φ(η) for with η for several values of Lewis number Le.

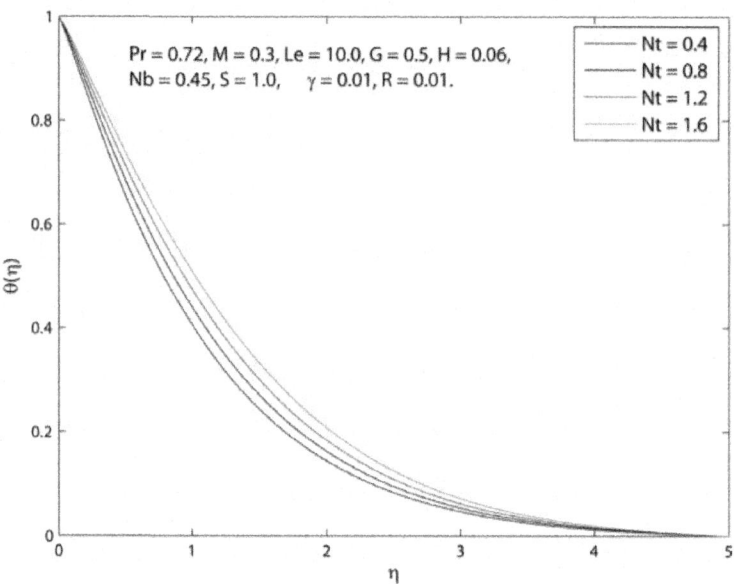

Figure 7: Variation of temperature $\theta(\eta)$ with ηseveral values of thermophoresis parameter Nt.

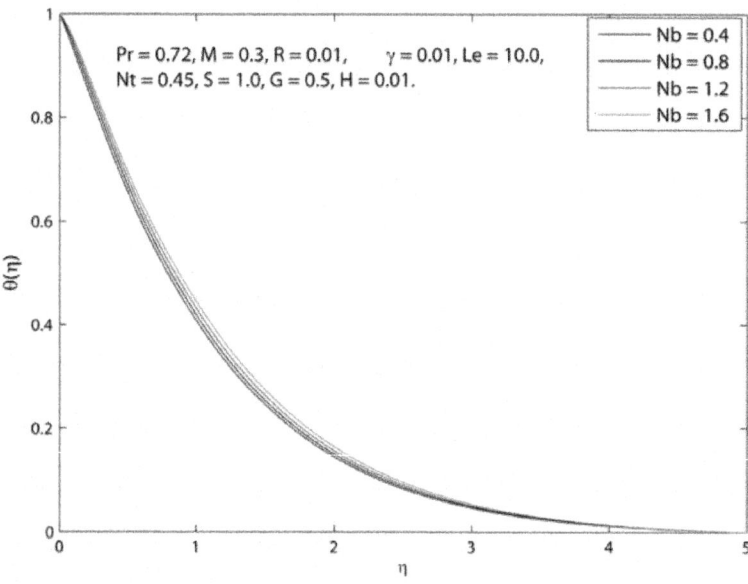

Figure 8: Variation of temperature $\theta(\eta)$ with η for several values of Brownian motion Nb.

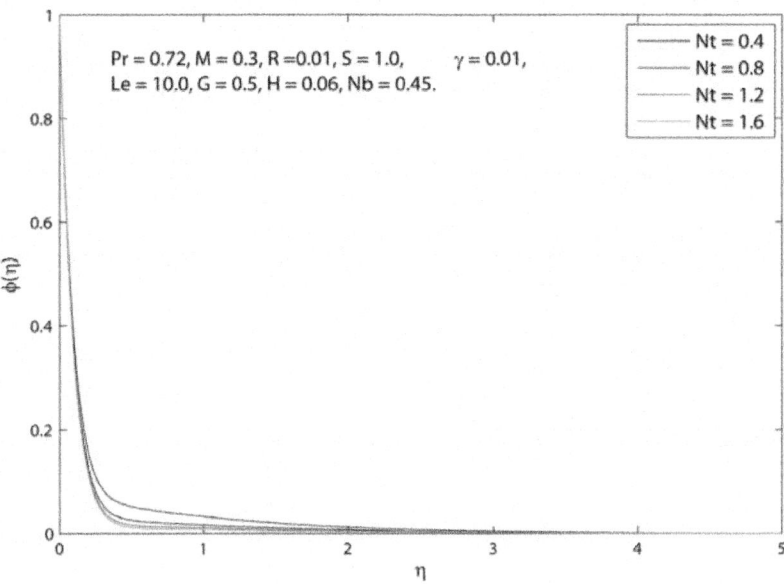

Figure 9: Variation of concentration of ϕ(η) with η for several values of thermophoresis parameter Nt.

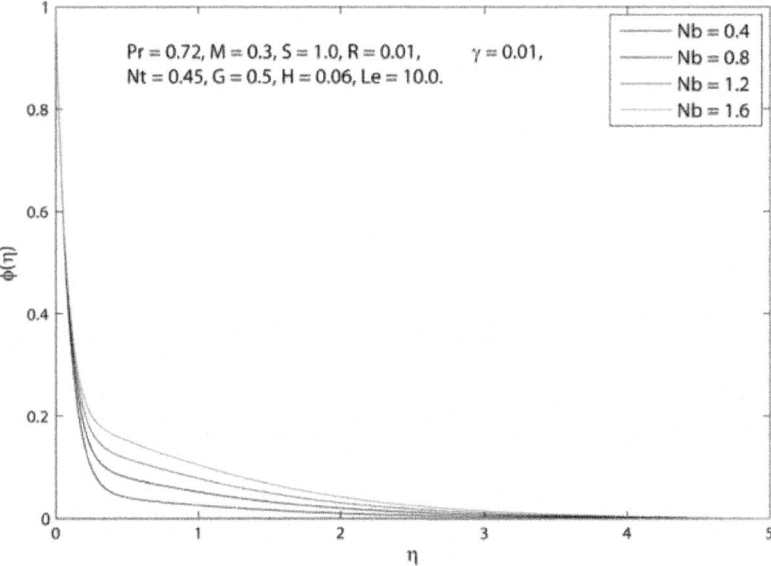

Figure 10: Variation of concentration of ϕ(η) with η for several values of Brownian motion parameter Nb.

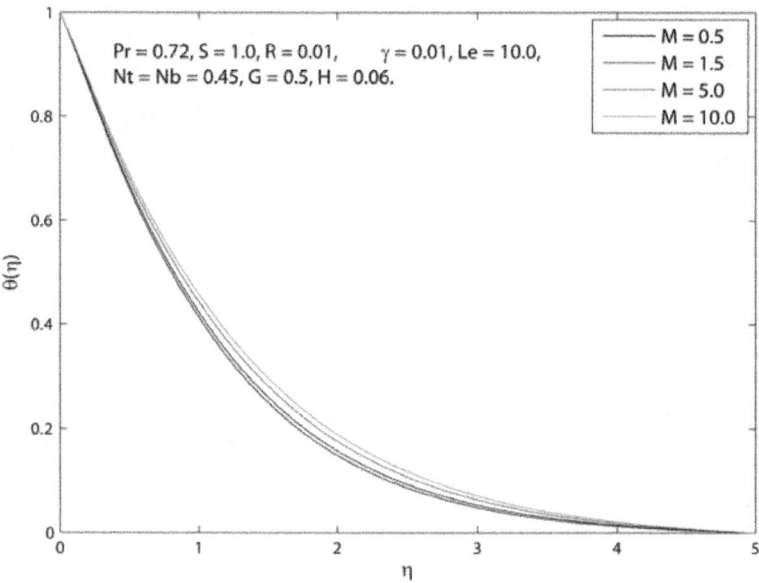

Figure 11: Variation of temperature θ(η) with η for several values of magnetic parameter M.

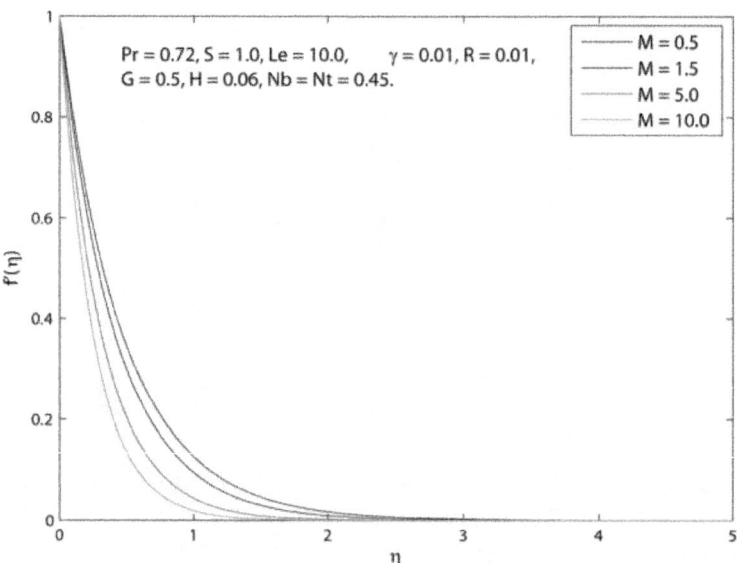

Figure 12: Variation of velocity $f'(\eta)$ with η with several values of Magnetic parameter M.

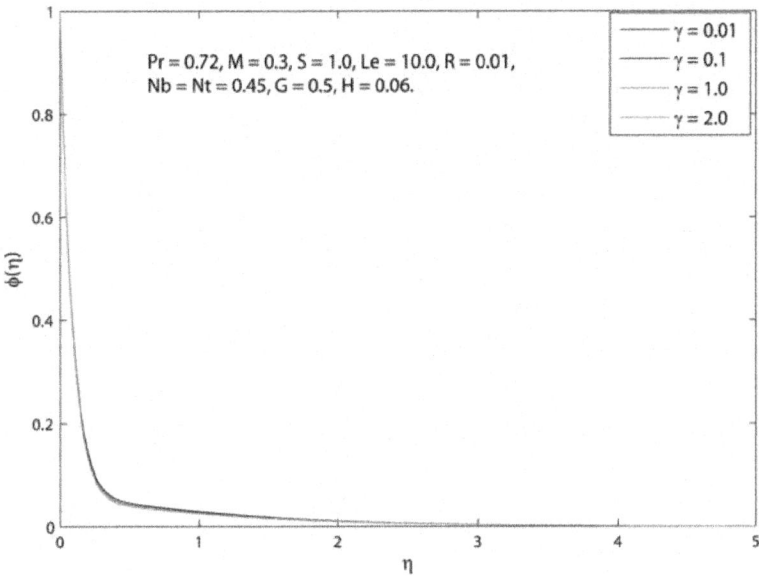

Figure 13: Variation of concentration ɸ(η) with η for several values of chemical reaction parameter γ.

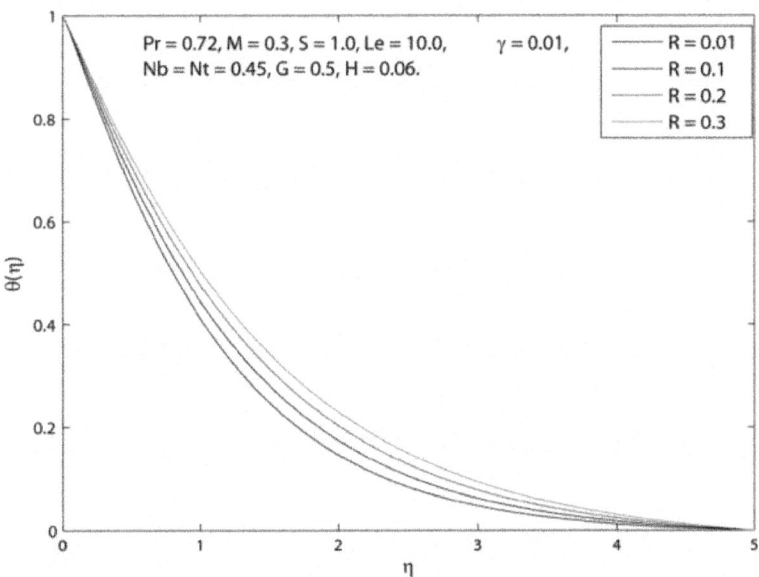

Figure 14: Variation of temperature θ(η) with η for several values of thermal radiation parameter R.

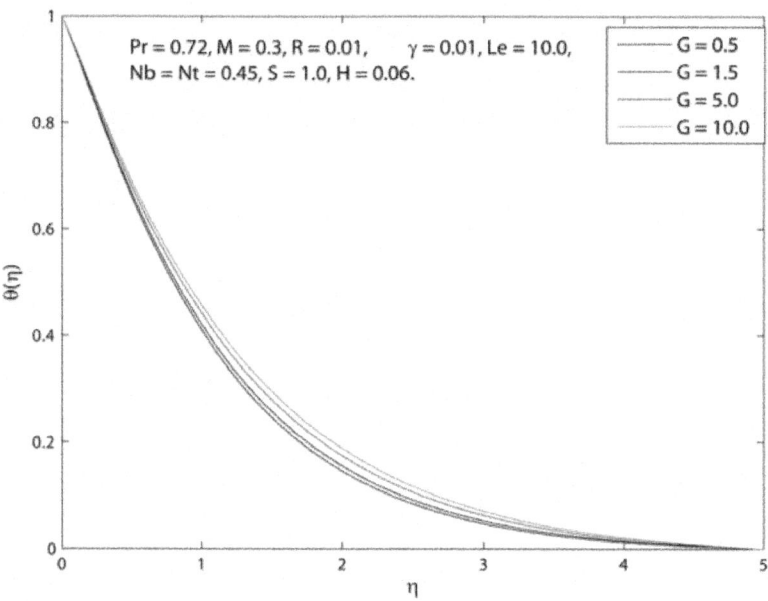

Figure 15: Variation of temperature θ(η) with η for several values of porosity parameter G.

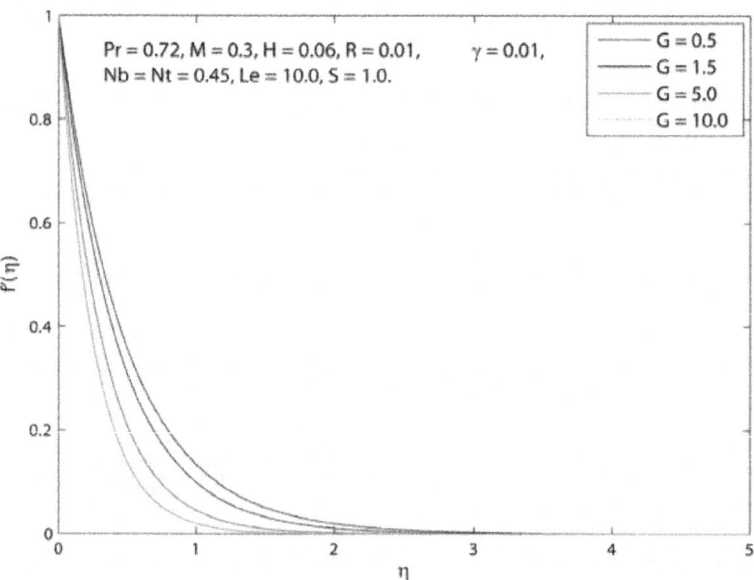

Figure 16: Variation of velocity $f'(\eta)$ with η for several values of porosity parameter G.

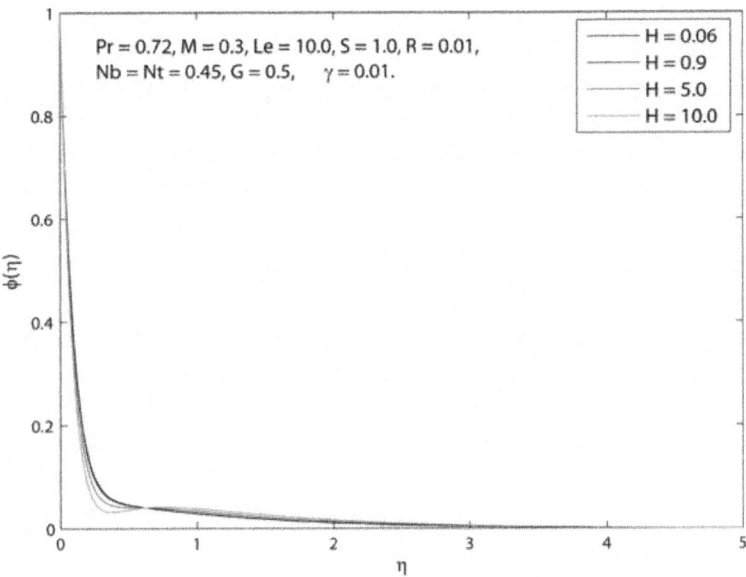

Figure 17: Variation of concentration (η) with η for several values of joule heating effect H.

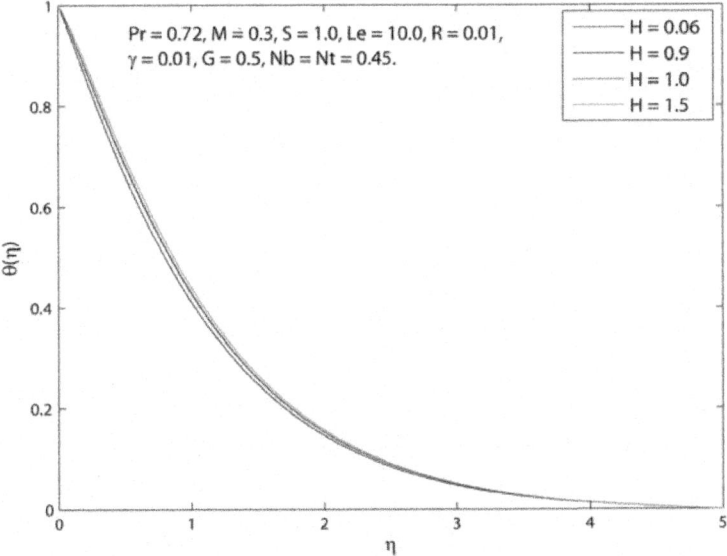

Figure 18: Variation of temperature θ(η) with η for several values of joule heating effect H.

CONCLUSIONS

A numerical study corresponding to the flow and heat transfer in a steady flow region of nanofluid over an exponential stretching surface and effects of chemical reaction, thermal radiation, magnetic, suction parameter, porosity parameter and joule heating parameters is examined and discussed in detail. The main observations of the present study are as follows.

An increase in suction parameter leads the velocity, temperature and concentration profiles to decrease. For larger values of Le suppress the concentration profile i.e. inhibit nanoparticle species diffusion, as observed. There will be a much greater reduction in the concentration boundary layer thickness. As Nt increases, temperature profile increases but the concentration profile decreases. With increasing values of Nb, both temperature and concentrations profiles increase. As increase in chemical reaction paramter γ leads the concentration profiles to decrease. The impact of porosity parameter shows that velocity profile is decreasing and temperature profile is increasing. This is because the porous medium inhibits the fluid not to move freely through the boundary layer. This leads the flow to increase thermal boundary layer thickness. Joule heating parameter reduces the temperature and concentration of nanofluid.

REFERENCES

1.   Choi, S.U.S. (1995) Enhancing Thermal Conductivity of Fluids with Nanoparticles. In: Siginer, D.A. and Wang, H.P., Eds., Developments and Applications of Non-Newtonian Flows, ASME, New York, 99-105.

2.   Eastman, J.A. and Choi, S.U.S. (1997) Enhanced Thermal Conductivity through Development of Nanofluids. MRS Proceedings, 457, 3-11. http://dx.doi.org/10.1557/proc-457-3

3.   Routbort, J.L., Yu, D.M. and Choi, S.U.S. (2008) Review and Comparison of Nanofluid Thermal Conductivity and Heat Transfer Enhancements. Heat Transfer Engineering, 29, 432-460. http://dx.doi.org/10.1080/01457630701850851

4.   Buongiorno, J. (2006) Convective Transport in Nanofluids. ASME Journal of Heat Transfer, 128, 240-250. http://dx.doi.org/10.1115/1.2150834

5.   Cheng, L. (2008) Nanofluid Two Phase Flow and Thermal Physics: A New Research Frontier of Nanotechnology and Its Challenges. Journal of Nanoscience and Nanotechnology, 8, 3315-3332. http://dx.doi.org/10.1166/jnn.2008.413

6.   Rajabpour, A. and Akizi, F.Y. (2013) Molecular Dynamics Simulation of the Specific Heat Capacity of Water-Cu Nanofluids. International Nano Letters, 3, 58. http://dx.doi.org/10.1186/2228-5326-3-58

7.   Hu, W. and Buongiorno, J. (2005) Nanofluid Coolants for Advanced Nuclear Power Plants. Proceedings of ICAPP'05, Seoul, May 2005, 15-19.

8.   Gupta, P.S. and Gupta, A.S. (1997) Heat and Mass Transfer on a Stretching Sheet with Suction or Blowing. The Canadian Journal of Chemical Engineering, 55, 744-746. http://dx.doi.org/10.1002/cjce.5450550619

9.   Emmanuel, S. and Khan, S.K. (2006) On Heat and Mass Transfer in a Viscoelastic Boundary Layer Flow over an Exponentially Stretching Sheet. International Journal of Thermal Sciences, 45, 819-828.

10. Kim, Y.J. (2000) Unsteady MHD Convective Heat Transfer Past a Semi-Infinite Vertical Porous Moving Plate with Variable Suction. International Journal of Engineering Science, 38, 833-845.

11. Subhas Abel, M. and Siddheshwar, P.G. (2007) Heat Transfer in a Viscoelastic Boundary Layer Flow over a Stretching Sheet with Viscous Dissipation and Non-Uniform Heat Source. International Journal of Heat and Mass Transfer, 50, 960-966. http://dx.doi.org/10.1016/j.ijheatmasstransfer.2006.08.010

12. Nadeem, S., Zaheer, S. and Fang, T.G. (2011) Effects of Thermal Radiation on the Boundary Layer Flow of a Jeffrey Fluid over an Exponentially Stretching Surface. Numerical Algorithms, 57, 187-205. http://dx.doi.org/10.1007/s11075-010-9423-8

13. Ishak, A. (2011) MHD Boundary Layer Flow Due to an Exponentially Stretching Sheet with Radiation Effect. Sains Malaysiana, 40, 391-395.

14. Al-odat, M.Q., Damseh, R.A. and Al-azab, T.A. (2006) Thermal Boundary Layer on an Exponentially Stretching Continuous Surface in the Presence of Magnetic Field Effect. International Journal of Applied Mechanics and Engineering, 11, 289-299.

15. Kumar, H. (2013) Heat Transfer in MHD Boundary Layer Flow through a Porous Medium, Due to a Non-Isothermal Stretching Sheet, with Suction, Radiation and Heat Annihilation. Chemical Engineering Communications, 200, 895-906. http://dx.doi.org/10.1080/00986445.2012.727509

16. Chand, G. and Jat, R.N. (2014) Flow and Heat Transfer over an Unsteady Stretching Surface in a Porous Medium. Thermal Energy and Power Engineering, 3, 266-272.

17. Sudhakar, K., Srinivas Raju, R. and Rangamma, M. (2013) Hall Effect on Unsteady MHD Flow Past along a Porous Flat Plate with Thermal Diffusion, Diffusion Thermo and Chemical Reaction. International Journal of Physical and Mathematical Sciences, 4, 370-395.

18. Subhas Abel, M., Kumar, K.A. and Ravi kumara, R. (2011) MHD Flow, and Heat Transfer with Effects of Buoyancy, Viscous and Joules Dissipation over a Nonlinear Vertical Stretching Porous Sheet with Partial Slip. Engineering, 3, 4.

19. Hamza, M.M., Isah, B.Y. and Usman, H. (2011) Unsteady Heat Transfer to MHD Oscillatory Flow through a Porous Medium under Slip Condition. International Journal of Computer Applications, 33, 266-272.

20. Sharma, P.R. and Singh, G. (2010) Effects of Variable Thermal Conductivity, Viscous Dissipation on Steady MHD Natural Convection Flow of Low Prandtl Fluid on an Inclined Porous Plate with Ohmic Heating. Meccanica, 45, 237-247.

21. Cebeci, T. and Bradshaw, P. (1988) Physical and Computational Aspects of Convective Heat Transfer. Springer, New York. http://dx.doi.org/10.1007/978-1-4612-3918-5

22. Rudraswamy, N.G. and Gireesha, B.J. (2014) Influence of Chemical Reaction and Thermal Radiation on MHD Boundary Layer Flow and Heat Transfer of a Nanofluid over an Exponentially Stretching Sheet. Journal of Applied Mathematics and Physics, 2, 24-32. http://dx.doi.org/10.4236/jamp.2014.22004

23. Mustafaa, M., Hayat, T. and Obaidat. S. (2013) Boundary Layer Flow of a Nanofluid over an Exponentially Stretching Sheet with Convective Boundary Conditions. International Journal of Numerical Methods for Heat and Fluid Flow, 23, 945-959. http://dx.doi.org/10.1108/HFF-09-2011-0179

Chapter 9

A HANDS-ON PROJECT-BASED MECHANICAL ENGINEERING DESIGN MODULE FOCUSING ON SUSTAINABILITY

Tom Joyce[1], Iain Evans[1], William Pallan[2], and Clare Hopkins[1]

[1]School of Mechanical and Systems Engineering, Newcastle University, Stephenson Building, Claremont Road, Newcastle upon Tyne NE1 7RU, UK

[2]WP Consulting, Newcastle upon Tyne, UK

ABSTRACT

Design is a crucial element of engineering education. The mode of learning design techniques and knowledge is key to the relationship that students gain with this subject. In the School of Mechanical and Systems Engineering at Newcastle University modules in Design and Manufacturing are taught in the first and second year. This paper outlines the transition of the Year 2 Design and Manufacturing module from a paper-based exercise to a design–build–test project-based group exercise over a period of three years between 2007/08 and 2009/10, using an iterative process of student feedback to advise this process of module change. Through this evaluation five key elements were identified, which inform the process of the introduction of practical group working in an engineering setting – these are presented as having utility for engineering education in general. Two conclusions derived from the analysis are also outlined, together with their implications for the engineering higher education sector.

INTRODUCTION

The School of Mechanical and Systems Engineering (MSE) at Newcastle University offers a three-year BEng degree in Mechanical Engineering together with a suite of four-year MEng degrees in Mechanical Engineering based disciplines. All degrees are accredited by the Institution of Mechanical Engineers or the Institution of Engineering and Technology. As with many such degrees in the UK, engineering design is seen as a core subject and is taught in MSE in both the first and second year.

A change in the staff involved in the module to include the first three authors offered an opportunity to reconsider both the pedagogical approaches and module content to expand the opportunities for project-based collaborative design team learning. This paper analyses the process of change that occurred over the three years during which this Stage 2 Design module became a Design and Manufacturing module, and completed the transition from a paper-based design exercise to a group exercise which required students to design, build and test a wind turbine within a £100 budget.

Teaching for the module took place over two terms and the module represented 20 credits of a total of 120 credits for the year. Four hours per week of contact time were allocated, broken down into a one-hour lecture and a three-hour slot where the students worked in groups. At this time group/team working was not standard practice for the programme. The assessment (and percentage marks) for academic year 2007–08 onwards consisted of a group essay related to sustainable development (5%), an interim report (33%), seven updates within which students were encouraged to reflect upon how their project learning linked to UK-SPEC requirements (26%) and a Final Report and Logbook assessment (36%).

By eliciting comprehensive feedback from students on their learning experiences within the module, it has been possible to engage in a reflective process which facilitated modifications and refinements as a response to this feedback. Detailed analyses of three consecutive sets of annual student feedback also allowed the identification of five key factors in the process of the development of project and team-based learning within this module. These key learning points were that 1) engineering students favour hands-on learning 2) seeing a project through to completion provides a sense of satisfaction/achievement 3) working as part of a team facilitates mutual support and collaborative learning 4) project 'authenticity' is required for full project engagement and 5) students have a dual need for autonomy and support, which may be challenging for academic staff to navigate. These key learning points will be expanded upon within the discussion section.

RATIONALE FOR THE INTRODUCTION OF PEDAGOGICAL CHANGE

As Professor Chris Pearce, a member of the Visiting Professor Team of the Royal Academy of Engineering, is quoted as saying of design teaching:

And what do we need to teach? We don't. We need to give the opportunity to gain experience and awareness in multidiscipline team environments and let the confidence of youth loose on a prepared world. What can we give students in a university department? Experience of working in multidisciplinary teams working on realistic projects. (Royal Academy of Engineering 2005.)

This understanding that students learn most effectively when they have opportunities for experiential, generative learning within groups and that group working represents a preparation for roles within the engineering profession (Springer *et al.* 1999) was fundamental to the redesign of this module. As Dym *et al.* (2005) suggest "Design is what engineers do, and the intelligent and thoughtful design of the engineering curriculum should be the community's first allegiance" (p114).

The literature around the use of problem-based and project-based learning within engineering education is often confused or conflated. Both approaches are based on self-direction and collaboration and both may be multi-disciplinary (Perrenet *et al.* 2000). Mills & Treagust (2003, p8) suggest that there are five essential differences between these two approaches. These are that, for project-based learning:

- Project tasks are closer to professional reality and therefore take place over a longer time-scale.
- Project work is focused on the *application* of knowledge whilst problem-based learning focuses on the *acquisition* of knowledge.
- Project-based learning is usually supported by subject teaching.
- Project teams have to manage time, resources and make decisions about task and role differentiation.
- Project teams have stronger self-direction.

Project-based learning employs an inductive, learner-centred approach to the acquisition of knowledge and skills. It is underpinned by constructivist theory, which assumes that all new understandings are built upon previous knowledge and that the interweaving of knowledge and practical skills can help students to consolidate their understandings (Tempelman & Pilot 2011). Project-based learning encourages students to learn together and problem-solve collectively.Elshorbagy & Schönwetter (2002) suggest that this process is, in itself, generative and that learning is reinforced when students teach each other. Project-based learning gives students the opportunity to develop their communication, problem-solving and team-working skills which will be relevant in their future careers (Elshorbagy & Schönwetter 2002). This approach is enhanced when projects are combined with challenge-based or enquiry-based learning (Bramhall *et al.* 2008, Powell *et al.*2008), when students learn collaboratively (Prince 2004) and when projects not only have a 'hands-on' practical focus, but students are also given greater responsibility for their own learning (Lambert *et al.* 2008).

Working together with other students as a team is a fundamental part of project-based work. Team working provides opportunities for peer-to-peer sharing of knowledge and skills, mutual academic support and enhanced collaborative learning (Joyce & Hopkins 2011). Teams can also represent a form of 'learning community'. Zhao & Kuh (2004) found learning communities were positively linked with academic performance, student engagement and perception of their college environment (p124). Similarly, Zepke & Leach (2010) write that students who learn together and from each other, who "make connections between ideas whilst drawing on the ideas, experiences and knowledge of others are most deeply engaged" (p172). Berglund (2012, p31) suggests that team climate is of great importance in project work and that where there is a strong sense of self-efficacy, this both strengthens achievement and promotes deep approaches to learning. Berglund (2012)also reports that students judge themselves more positively where there is ease of communication and proximity between group members. When students have not had experiences of either practical working or working as part of a group of peers within their university education, they may have difficulty in making the transition between university and industry, as reported by Flores (2012) in his study of Mechanical Engineering students in Portugal.

Lecturers who facilitate project-based learning are required to adopt a very different role from that of information provider employing a traditional lecture format and deductive approach with a large number of students. They need skills in facilitating small group learning (Perrenet *et al.* 2000), to be able to help students to bridge the gap between theoretical knowledge and real-life problem solving (Elshorbagy & Schönwetter 2002), and to guide students through their mistakes to a greater sense of design task-specific mastery (Carberry *et al.* 2010). Groups who perceive their supervisors to be readily available, actively-involved and reassuring were found by Berglund (2012, p31) to function most effectively. Lecturers who are also experienced professional design engineers will be uniquely placed to offer support and supervision to students (Hayhurst *et al.* 2012).

It is recognised that the motivation of students to learn is increased when they perceive themselves to be developing the professional skills needed for their future careers (Fang 2012). If students are to engage in a meaningful way with their project then it is essential that it contains a recursive interweaving of theory and practice. 'Authentic' projects provide a meaningful context for student learning and consequently stimulate development of expertise and engagement with the project (Tempelman & Pilot 2011, Fang 2012). Project work that is structured to reflect the real-world requirements of professional engineering design facilitates the development of skills such as professional report writing (Dym 1994).Christiaans & Venselaar (2005) for example, found that students engaged in experiential design projects acquired "general process knowledge, heuristics and higher-order rules that can be transferred to a wide variety of design situations" (p226).

In the second year Design module it was felt that the introduction of a group-oriented, project-based approach would assist students with the transition from a purely theoretical to a practice-based framework (Mills & Treagust 2003). It was hypothesised that this change would also provide the opportunity for students to develop the skills and attributes of "integrity, independence, impartiality, responsibility and competence and frequently discretion" required for sound decision making outlined by the Royal Academy of Engineering (2005, p1).

EVALUATION METHODOLOGY

Contemporaneous with the introduction of modifications and changes to the module, a revised ten-item feedback questionnaire was introduced (Appendix 1). The form consisted of ten questions, each of which invited students to anonymously choose a response on a five-point scale. The questions covered the structure of the module, the input of teaching staff and questions specific to the student completing it. In addition, the questionnaire offered the opportunity to add freehand comments. Students were invited to describe two good features of the module and to make two suggestions for improvement, as well as being free to make any other comments about the module.

The standard student evaluation is often criticised for being insufficiently sensitive to allow students to provide feedback related to the minutiae of a specific module (Huxham *et al.* 2008). Doubts are often also expressed that the data collected is used as a means of improving the quality of teaching and learning (Kember *et al.* 2002). Students providing feedback on this module were made aware that their ratings and comments would be used to make refinements to its structure and the teaching and learning strategies employed as the recommendations of each cohort were integrated into the module design for forthcoming years.

Analysis of the survey took place as each group of students neared the end of the module. In each case the ratings data were entered into a spreadsheet and graphical representations of the data were created. The freehand qualitative comments were analysed for themes. The list of themes was read repeatedly to gain a sense of their content and a set of categories were generated from these readings. Each comment was then tested against the categories; it was possible for each comment to be matched against a category with the exception of a small number of outlying comments. This method is a simplified version of the constant comparative method (Boeije 2002). Once the categories had been assembled it was possible to create meaningful sub-categories which allowed the nuances of feedback to become visible and coordinated.

THE PROCESS OF CHANGE

In addition to being responsive to student feedback, module design was also influenced by the first three authors' previous experience of teaching the module together with their reflective discussions. Table 1 summarises developments over the three academic years.

Table 1: Evolution of the Design and Manufacturing module over three years

Year	Size of Stage 2 cohort	Project group size	Module format	Assessment methodology	Changes and modifications based on student feedback and staff initiatives
2007/08 Year 1 (15 credits)	63	7	**Project title**: 'Greening our Homes' Collaborative group project. Develop three conceptual designs of energy-saving/ generation devices in 1st term. Finalise design of one concept in second term.	Seven 'weekly updates'. Completion of individual logbook. 3000 word group interim and final reports. Group interim and final presentations.	N/A
2008/09 Year 2 (20 credits)	73	7 or 8	**Project title**: 'From Kilobytes to Kilowatts' Collaborative group project. Design/build/test a wind turbine. Term 1 – given redundant computer and printer to design and manufacture turbine (tested end of term). Term 2 – budget of £100 amend original design and manufacture (basic hand tools). Use of wind tunnel for testing.	Group essay on sustainability. Seven 'weekly updates'. Completion of individual logbook. 3000 word group interim and final reports – groups decide allocation of marks. Assessment of wind turbines at end of first and second terms. Group interim and final presentations.	Provision of workshop space and basic hand tools to allow hands-on learning. Allocation of £100 budget for locally sourced material and components in second term. Reduction in number of presentations and option to submit DVD instead of presentation. Only one submission of logbook. Every group has same motor.
2009/10 Year 3 (20 credits)	95	7 or 8	**Project title**: 'From Kilobytes to Kilowatts' Collaborative group project. (Initial project using redundant computer/ printer restricted to three weeks). £100 budget available from week three. Design/build/test a wind turbine (as in 2008/09). Use of wind tunnel for testing process.	Group essay on sustainability. Seven 'weekly updates'. Completion of individual logbook. 3000 word group interim and final reports – groups decide allocation of marks(option to submit DVD). Assessment of wind turbines at end of first and second terms.	Computer/printer exercise reduced duration (3 weeks rather than 1 term). No longer required to generate power. Remainder of time spent on 'final' design and budget made available after three weeks rather than in second term.

Findings

The quantitative data collected from relevant parts of the feedback questionnaire are presented in graph form in order to show a direct comparison across years. The questionnaires were completed by 52 students (83% of group) in 2007/08, by 57 students (78% of group) in 2008/09 and 82 students (86% of group) in 2009/10. In total there were 191 data sets gathered over a three-year period.

It must be stressed that the questionnaire format elicited quantitative data through a structured format, whilst no such format was provided for qualitative data (Appendix 1). It was not unusual for respondents' statements to relate to multiple aspects of the module, increasing the total number of items in each category.

Qualitative feedback from the questionnaires over the three years reflected the increasingly hands-on and practical nature of the module, but the key positive themes remained constant. These were that it represented:

- A practical opportunity to gain skills in preparation for professional practice.
- The freedom for creativity and innovation; the opportunity to 'learn by doing' and to 'see a project through'.
- A group-work experience which was welcomed even when difficulties were encountered.

Requests for module modifications similarly altered as the module changed, although increasingly revealed respondents' engagement with the practical nature of their project. At the end of the 2007/08 academic year the main change required was the opportunity to manufacture a design rather than to stop once engineering drawings had been produced. In 2008/09 the focus changed to concerns about group functioning and the need for more workshop time. By 2009/10 the feedback had moved to requests for specific lectures or knowledge such as "more lectures on aerodynamics". This latter comment revealed an interesting difference in approach. Many students seemed keen to maximise power (group competition to obtain the highest Watt output per pound spent was strong) and craved an imaginary 'optimum' blade profile to give them this. Staff opinion, tempered by industrial experience, was that such concerns were not the ultimate priority in a device with a 25-year intended lifespan. The requests from students for more workshop time indicated a preference for hands-on work and it also reinforced the need for good time management skills. In response to other student requests, additional hand tools were made available in 2008/09 and each team was provided with an identical motor to reduce variability across projects.

The quantitative and qualitative questionnaire elements have been interwoven into a narrative around the key themes which are: (a) students' perceptions about the workload of the module (b) their experiences of group work (c) the teaching environment and (d) the module content as a preparation for professional practice.

Perceptions of Workload

Across the three years of module development, feedback indicated a growing perception of increasing workload; the number ticking the category of 'heavy' almost doubled from 27% in 2007/08 to 52% in 2009/10 (Figure 1). Paradoxically this shift in perception occurred despite staff efforts to reduce the workload and to simplify the structure and assessment procedures for the module. Students reported spending a higher number of hours outside class on study related to the module, rising from 21% who dedicated more than six hours each week to out-of-class work on the project in 2007/8, to 44% dedicating the same amount of time in 2009/10. By 2009/10, a further 12% were spending more than eight hours per week on module work (Figure 2). However, when a calculation was done of how many hours the students had spent on the module (based on their own perception) this actually matched the hours expected by the university.

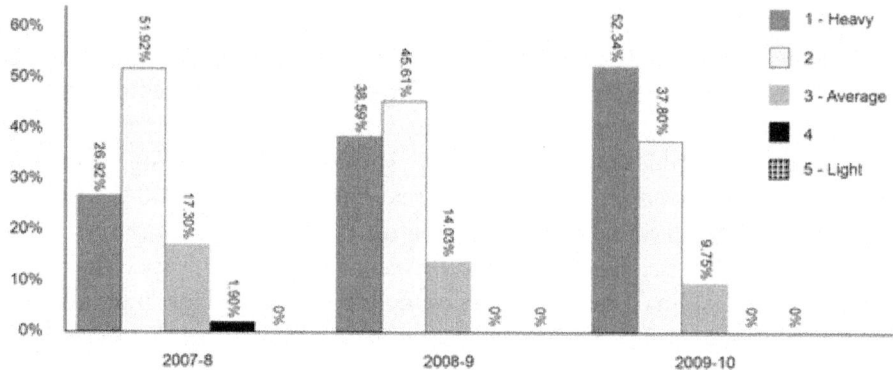

Figure 1: Perception of weight of workload relative to other modules

Interestingly, qualitative feedback throughout the three years focused not on students' perceptions of the number of hours spent completing their work, but on the positive aspects of having a prolonged period of time to work on a project. In 2007/08 one student commented that he had become very involved in the project because of its length; in 2008/09 another respondent wrote that "It took up a lot of time but was fun and we learnt a lot". In the same year, suggestions for change appealed for workshops to be available for more than three hours a week and for "more access to turbines out of lab hours". Feedback in 2009/10 included multiple requests for "greater access to workshops outside of working time" demonstrating students' engagement with their project. This sense of deep engagement is also revealed in respondents' requests for 'more time' for various aspects of work, such as "more time to carry out research", "more time to work as a group", "more time in labs and more time for testing", "more testing time/time to adjust, improve, then test again (with real wind tunnel)".

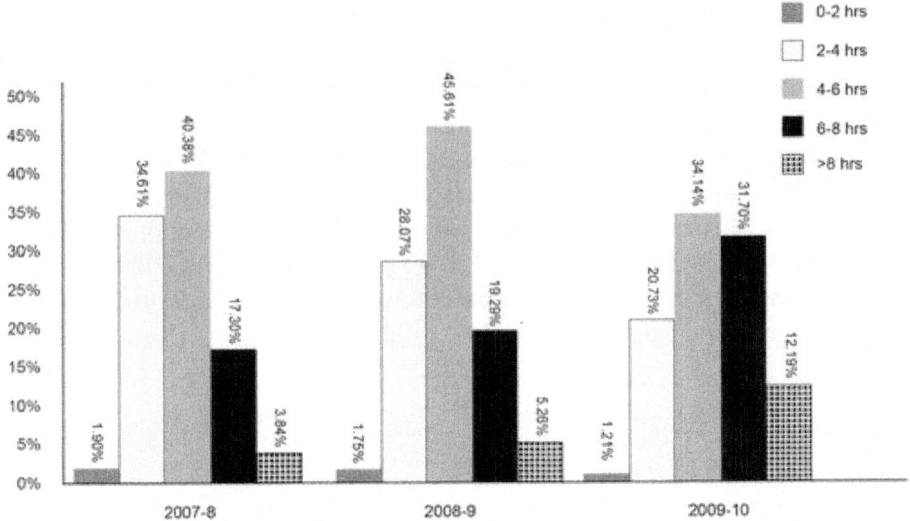

Figure 2: Hours worked per week outside class

Group Working

Group working appeared to enhance the students' sense of ownership of their project and their learning. In 2007/08 all groups had seven students but, as student numbers increased, group size occasionally increased to eight due to practical considerations such as availability of workshop space and worktables. It was therefore very difficult to reduce group size despite qualitative feedback indicating that larger groups could experience problems in maintaining communication and allocating work equitably.

Despite such concerns, the development of team working and communication skills was highly prized by each year group as a preparation for professional practice. Among the response to the open-ended question: "give two good features of the module", in 2007/08 team working was mentioned positively by 15 (29%) survey respondents, by 23 (40%) in 2008/09 and by 28 (34%) in 2009/10. The team experience was valued because it provided an experience of working, as one student put it, "in teams and under pressure", others commented that it represented "actual practice on project management", a "great team working exercise – very valuable skills learned" and another wrote that it represented "learning from each other, different people have different understandings".

Teaching Environment

Throughout the three years of feedback there was also an increasingly positive rating of the teaching environment. Those rating it in the two highest categories rose from 72% in 2007/08 to 82% in 2008/09 and to 87% in 2009/10 (Figure 3). In parallel with this the number of responses rating lecturers' interest and enthusiasm for the subject in the two highest categories stayed high across all three years, varying only between 94% and 96% (Figure 4). This positive response is also reflected in the students' rating of the overall teaching on the module, which rose from 72% in the two highest ratings on the scale in 2007/08 to 91% in 2008/09 and only slightly lower at 90% in 2009/10 (Figure 5). A comment from a student in 2007/08 was that there was "good structure for both semesters. Limitless concepts could be used".

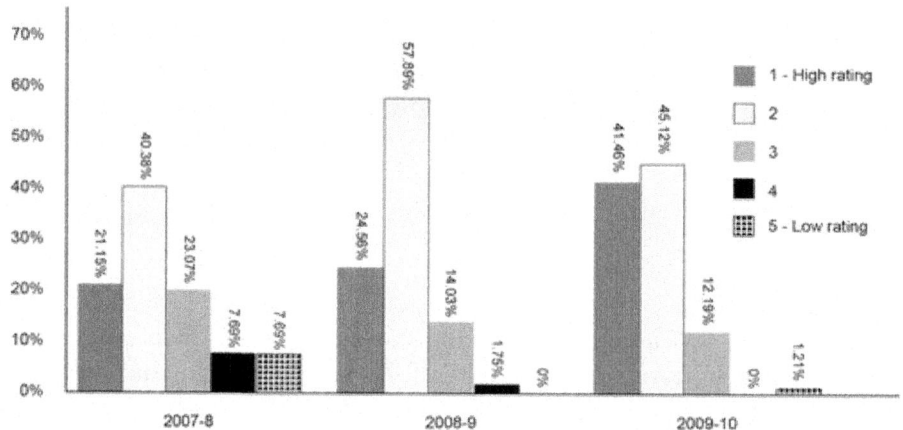

Figure 3: Perception of the teaching environment

The availability of help, support and feedback from staff was noted as important within all qualitative feedback, both when it was seen as being sufficient and also when additional support was felt to be necessary. Students appeared to enjoy the freedom that the open-ended design projects gave them, but they also wanted to be able to call upon a great deal of staff support. There was recognition that the industrial experience of teaching staff was valuable and that, as one student expressed it, a "professional approach by lecturers inspires a professional output by students".

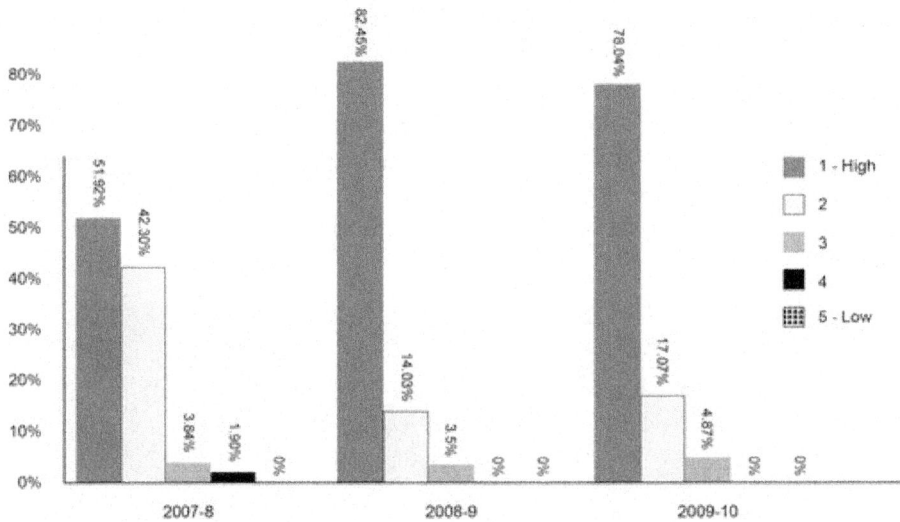

Figure 4: Lecturers interest and enthusiasm for their subject

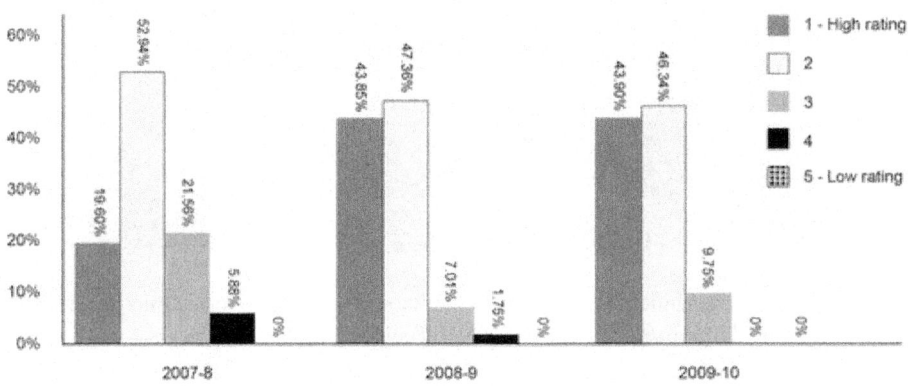

Figure 5: Overall rating of teaching on the module

Module Content

Module content was recognised as highly significant across all three years. The major factor attracting positive feedback was its relevance to students' professional skills development. The module was seen as "practical", "hands on" and linking theory to engineering practice. These comments were made by 28 (54%) respondents in 2007/08, 51 (89%) in 2008/9 and 36 (44%) in 2009/10. One student's words exemplify this understanding when he wrote that the module "brought ALL aspects of engineering together and showed the relevance of each". Other comments show that the module presented an opportunity to "learn by doing", that it linked theory and practice because "we actually got to do stuff"; that it was a "practical course – more enjoyable/learn more", but especially that it "helps prepare for real-world design problems". The acquisition of new skills such as project management, report writing and budget management were commented on by respondents in each group (eight (15%) in 2007/08, 16 (28%) in 2008/09 and 17 (21%) in 2009/10). One of the final wind turbine designs manufactured on the £100 budget is shown in Figure 6.

Figure 6: A wind turbine based on components and materials from a maximum expenditure of **£100** (large motor at rear supplied free of charge)

The reasons for the module providing this positively rated experience were multi-factorial. The descriptions which appeared most frequently amongst qualitative feedback were "freedom" and "independence" and the linked, resultant concepts of "creativity" and "innovation" (mentioned collectively ten times (19%) in 2007/08, 11 times (19%) in 2008/09 and 21 times (26%) in 2009/10), becoming more prominent as the module changed from design only. Respondents' comments on these linked concepts were that the module had provided; "freedom for creativity", "helps with thinking and being inventive", "freedom to implement our ideas" and "freedom to design creatively".

Another positive aspect of the module was that it promoted continuity and a sense of completion. One student commented that it allowed her to "get to see the whole life cycle of the project. Made me proud to see something at the end". This was highly valued as it promoted learning through being "given the chance to learn by fixing our mistakes" or as another student put it succinctly "made u think – if it didn't work, made u rethink". By 2008/09 26% of respondents commented on the importance of this complete life cycle aspect of the project.

The sustainability topic was very important too – the choice of a renewable energy project further enhanced students' perception of it as 'authentic' and increased their interest. Respondents reported being able to integrate and use knowledge gained across their degree programme in this design project work and also indicated that they had acquired new process abilities such as time-management, team working, organisational and communication skills.

In Summary

The experience of working in teams and under pressure whilst managing a budget developed students' independence and resourcefulness and enhanced their sense of freedom and creativity. They valued the experience of simultaneously being allowed this high degree of freedom whilst receiving support from lecturers. It seemed that the freedom of the learning environment which fostered group and individual creativity, coupled with the opportunity to acquire professional skills, had more power to enhance student commitment than their perception of a heavy workload had to discourage them. This suggests that engineering students relish these challenges if they can be seen in the context of skills progression and preparation for professional practice. Students indicated this when they wrote of the module: "This is the exact thing that I expected to do on this course" and another that "I have learned more in this module than in the whole first year".

KEY LEARNING POINTS FROM EVALUATION OF THE MODULE

The process of module evolution and development has highlighted five key learning points which possibly have general applicability across a variety of modules where project-based learning and team working are being introduced into engineering programmes. These are outlined below.

Engineering students display a strong orientation towards learning in a 'hands on' or practical way and see this as being valuable preparation for their future careers.

Students in this study displayed a very high level of commitment to their project despite their perception that it represented a 'heavy' workload. Kember *et al.* (1996) found that students' perception of their workload may not be directly related to the actual hours worked, but may be multi-factorial and related to students' interest in the subject, the volume of content covered, the motivational approach used by lecturers, but most importantly, students' individual learning approaches. For many students in this evaluation the Design and Manufacturing module may have represented their first experience of group and project work and will therefore have influenced their perceptions of its workload. Students indicated that the work felt like a preparation for the 'real world', something which previous studies of engineering students' motivations have identified (Fang 2012, Tempelman & Pilot 2011). Allie *et al.* (2009) describe this process as developing discursive engineering identities, not simply acquiring knowledge, but a negotiation of a visible self: "engaging with engineering is an act that has implications for how others will see you" (p361).

The process of preparation for professional practice and the acquisition of transferrable skills, engineering language and terminology were frequently commented upon positively. Many students feel unprepared for this transition even if they have acquired the requisite knowledge (Flores 2012). Dahlgren *et al.* (2006, p583) comment that "the contextualisation of knowledge of working life occurs, if at all, late in the programme, or is left to the novices to handle individually". Having the opportunity to try out skills (in this case in the second year) allowed students to try out engineering identities within the safety of the institution and to gain confidence through practising this role.

The ability to see a process through from start to finish provides students with a sense of satisfaction and achievement

Being part of a process from concept through to design, manufacturing and testing was seen as a crucial part of the enjoyment and learning from the project. This feeling is exemplified in the words of one student who wrote in his logbook "nothing is more rewarding than to see what you've created in actual use, and to see it working is overwhelming". Incorporated within this sense of achievement are the circular, collaborative processes of having conceptualised a design framework within limited resources, negotiated manufacturing difficulties with a constrained supply of tools, gone through the process of testing, found solutions to problems encountered and brought forth a product for evaluation. Students who only encounter well-structured problems within a learning environment may not automatically be able to transfer their acquired problem-solving skills to an industrial environment where they meet real-life and therefore ill-structured problems (Jonassen *et al.* 2006).

Working together as a group/team provides opportunities for mutually supportive relationships where collaborative learning can take place

The ability to work as part of a team, especially a multi-disciplinary engineering team, is now an essential skill for all engineering graduates being demanded by industry (Royal Academy of Engineering 2010). Team working in an engineering educational setting can offer the opportunity for both mutual support and an environment in which students share knowledge and learn from each other (Kamsah & Talib 1995, Joyce & Hopkins 2011). It has also been found to help promote academic achievement as well as more favourable attitudes towards learning and persistence (Springer *et al.* 1999). Asking students to work together in groups is underpinned by the idea that learning takes place within social contexts and is achieved through sharing and the co-construction of meaning and understanding (Vygotsky 1978). It is not only the less able group members receiving instruction who benefit from the interaction; the team member who is sharing knowledge or skills also strengthens their own understanding within the process (Stump *et al.* 2011).

Working with a group of colleagues allowed students to move from design to finished product and to amalgamate the sum of their previous problem-solving experiences and differing perspectives, and bring these to bear on the design problems they faced, addressing both possibilities and constraints. As Jonassen *et al.* (2006, p144) suggest, "knowledge exists not only in the heads of learners, but also in the conversations and social relations among collaborators". Problem solving in this context becomes an iterative process where learning occurs when the contrast between new and previous experiences are integrated and become part of new understandings (Daly *et al.* 2012). Seeing a project through to its completion as a collaborative venture was viewed by students as both satisfying and another indication of gaining real-world expertise.

Difficulties arise in teams when they experience conflict or include members who fail to participate. When this occurs students may need staff support to overcome these problems (Parsons & Drew 1996, Burgland 2012).

If students are to engage effectively with the project they must view it as being relevant and 'authentic'.

For engineering students to engage enthusiastically and effectively with projects the projects must have relevance and the potential to provide a range of learning applicable for future career development (Tempelman & Pilot 2011, Fang 2012). In this evaluation the project with which students engaged most thoroughly and were prepared to devote most time to had sustainability at its heart and included recycled materials. UNESCO (2005, p9) recommend the acquisition of "skills, capacities, values and knowledge required to ensure sustainable development" at all levels of education. In addition, projects that incorporate interdisciplinary perspectives provide multiple opportunities for further discovery, innovation and accumulation of knowledge capital (Hayhurst *et al.* 2012).

There is a delicate tension between students' wish for autonomy and freedom and their wish for support

In this evaluation this delicate balance was clearly articulated within feedback and perhaps reflects the difficulty for academic staff in making judgements about how best to promote autonomy in learning groups. Dym (2005, p104) lists a series of skills and abilities associated with good designers as being able to 1) tolerate ambiguity that shows up in viewing design as inquiry or as an iterative loop of divergent–convergent thinking 2) maintain sight of the big picture by including systems thinking and systems design 3) handle uncertainty 4) make decisions 5) think as part of a team in a social process and 6) think and communicate in the several languages of design. Students in the second year of their programme may not have yet become comfortable with ambiguity and uncertainty and may require support which helps them to acquire these and the other skills described.

Tempelman & Pilot (2011) suggest that academic staff taking on the role of support and mentorship to students engaged on project-based work may be working outside their accustomed comfort-zone and may also require a level of support. They may be asked to examine in detail their theoretical knowledge base and be flexible in its application to practice in new and unexpected ways. This delicate balance is one which needs to be carefully negotiated and addressed.

CONCLUSIONS AND IMPLICATIONS

This evaluation of the transition of a second year Design and Manufacturing module from paper-based to a practical, hands-on, design–build–test exercise has revealed two conclusions which the authors suggest may have implications for the wider engineering higher education sector.

The first is that many of the things that students appreciate about project-based group working, such as seeing a project through from start to finish and working and learning with others on 'sustainable' and 'real world' projects, are all related to preparation for professional practice and employment. Many of those involved in the education of engineers will view this finding as confirmatory of what is already known.

What is perhaps more surprising is the second conclusion that students appear to be able to hold two contradictory positions simultaneously, without apparent awareness of any contradiction. Each succeeding group of students reported that whilst their perception that the module was increasingly demanding of their time and effort, they also felt a growing contentment with its content and teaching. Similarly, an increasing number of students throughout the three years highlighted both their enjoyment of the freedom/independence the module represented whilst simultaneously expressing a wish for a high level of staff support, monitoring and guidance.

At a time when engineering student numbers are expanding and higher education is faced with the increasingly complex challenge of ensuring that students graduate in a state of readiness to take up their place in industry, it is suggested that students undertaking modules featuring project and group work should be alerted to the possibility that they may experience contradictory perceptions and feelings as they negotiate their competing needs for independence to be creative and for academic support. Theorising this process as a stage on a developmental continuum towards a professional engineering identity could heighten students' ability to remain self-reflective as they go through this process of academic adjustment to full engineering identity.

APPENDIX 1: DESIGN AND MANUFACTURING II – FEEDBACK QUESTIONNAIRE

The purpose of this anonymous questionnaire is to provide feedback on the MEC2007 Design and Manufacturing II module. It aims to find out what went well and what did not. Please rate the module according to each of the criteria below. For each question in the table, please shade one circle with a pen or pencil. When you have done this, please answer the other questions. Your assistance is greatly appreciated.

1. The proportion of classes you attended	Nearly all		Half		Very few
	O	O	O	O	O
2. How well you felt the module was structured	High rating				Low rating
	O	O	O	O	O
3. The difficulty of the module relative to others	Too hard		Right		Too easy
	O	O	O	O	O
4. The course materials (handouts etc)	High rating				Low rating
	O	O	O	O	O
5. The workload of the module relative to others	Heavy		Average		Light
	O	O	O	O	O
6. The teaching environment	High rating				Low rating
	O	O	O	O	O
7. The lecturers' interest and enthusiasm for the subject	High				Low
	O	O	O	O	O
8. Your overall rating of teaching on the module	High rating				Low rating
	O	O	O	O	O
9. How many hours per week have you been working on this module outside of classes?	0–2	2–4	4–6	6–8	>8
	O	O	O	O	O
10. How available was help when you needed it?	Hardly ever				Nearly always
	O	O	O	O	O

Give two good features of this module and two suggestions for improvement

First good feature:

Second good feature:

Suggestion one:

Suggestion two

Feel free to make any further comments you wish to add about the module

REFERENCES

1.   Allie, S., Armien, M., Burgoyne, N. *et al.* (2009) Learning as acquiring a discursive identity through participation in a community: improving student learning in engineering education. European Journal of Engineering Education 34 (4), 359–367.

2.   Berglund, A. (2012) Do we facilitate an innovative learning environment? Student efficacy in two engineering design projects. Global Journal of Engineering Education 14 (1), 27–33.

3.   Boeije, H. (2002) A purposeful approach to the constant comparative method in the analysis of qualitative interviews. Quality and Quantity 36, 391–409. ,

4.   Bramhall, M., Radley, K. and Metcalf, J.E.P. (2008) Users as producers: students using video to develop learner autonomy. In Engineering Education Conference 14–16 July 2008. Loughborough, UK.

5.   Carberry, A., Lee, H.-S. and Ohland, M. (2010) Measuring engineering design self-efficacy. Journal of Engineering Education 99 (1), 71–79. ,

6.   Christiaans, H. and Venselaar, K. (2005) Creativity in design engineering and the role of knowledge: modelling the expert. International Journal of Technology and Design Education 15, 217–236. ,

7.   Dahlgren, M., Hult, H. and Dahlgren, L. (2006) From senior student to novice worker: learning trajectories in political science, psychology and mechanical engineering. Studies in Higher Education 31 (5), 569–586. ,

8.   Daly, S., Adams, R., Bodner, G. (2012) What does it mean to design? A qualitative investigation of design professionals' experiences. Journal of Engineering Education 101 (2), 187–219. ,

9.   Dym, C. (1994) Teaching design to freshmen: style and content. Journal of Engineering Education 84 (4), 303–310.

10. Dym, C., Agogino, A., Eris, O., Frey, D. and Leifer, L. (2005) Engineering design thinking, teaching and learning. Journal of Engineering Education 94 (1), 103–120. ,

11. Elshorbagy, A. and Schönwetter, D. (2002) Engineer morphing: bridging the gap between classroom teaching and the engineering profession. International Journal of Engineering Education 18 (3), 295–300.

12. Fang, N. (2012) Improving engineering students' technical and professional skills through project-based active and collaborative learning. International Journal of Engineering Education 28 (1), 26–36.

13. Flores, P. (2012) How do mechanical engineering students see their training and learning at university? Findings from a case study. Global Journal of Engineering Education 14 (2), 189–195.

14. Hayhurst, D., Kedward, K., Soh, H. and Turner, K. (2012) Innovation-led multi-disciplinary undergraduate design teaching. Journal of Engineering Design 23 (3), 159–184. ,

15. Huxham, M., Laybourn, P., Cairncross, S., Gray, M., Brown, N., Goldfinch, J. and Earl, S. (2008) Collecting student feedback: a comparison of questionnaire and other methods. Assessment & Evaluation in Higher Education 33 (6), 675–686. ,

16. Jonassen, D., Strobel, J., Beng Lee, C. (2006) Everyday problem solving in engineering: lessons for engineering educators. Journal of Engineering Education 95 (2), 139–151. ,

17. Joyce, T. and Hopkins, C. (2011) Working together: the positive effects of introducing formal teams in a first year engineering degree. Engineering Education 6 (1), 21–30.

18. Kamzah, M. and Talib, R. (1995) Assessing groupwork activities in engineering education. Assessment & Evaluation in Higher Education 20 (3), 289–300.

19. Kember, D., NG, S., Harrison, T., Wong, E. and Pomfret, M. (1996) An examination of the interrelationship between workload, study time, learning approaches and academic outcome. Studies in Higher Education 21 (3), 347–358. ,

20. Kember, D., Leung, D. and Kwan, K.P. (2002) Does the use of student feedback questionnaires improve the overall quality of teaching? Assessment & Evaluation in Higher Education 27 (5), 411–425.

21. Lambert, C., Basini, M. and Hargrave, S. (2008) The activity led learning within aerospace at Coventry University. In Engineering Education 14–16 July 2008. Loughborough, UK.

22. Mills, J. and Treagust, D. (2003) Engineering Education – is problem-based or project-based learning the answer? Australasian Journal of Engineering Education online publication 2003–04. Available athttp://www.aaee.com.au/journal/2003/mills_treagust03.pdf (accessed 4 November 2011).

23. Parsons, E. and Drew, S. (1996) Designing group project work to enhance learning: key elements. Teaching in Higher Education 1 (1), 65–80.

24. Perrenet, J., Boujhuijs, P. and Smits, J. (2000) The suitability of problem-based learning for engineering education: theory and practice. Teaching in Higher Education 5 (3), 345–358.

25. Powell, N., van Silfhout, R. and Hicks, P. (2008) Using enquiry-based learning (EBL) to prepare students for group work: lessons from successive implementations. In Engineering Education 14–16 July 2008. Loughborough, UK. Available athttp://www.heacademy.ac.uk/assets/documents/subjects/engineering/EE2008/p011-powell.pdf (accessed 5 June 2013).

26. Prince, M. (2004) Does active learning work? A review of the research. Journal of Engineering Education 93 (4), 223–231.

27. Royal Academy of Engineering (2005) Educating Engineers in Design. Available athttp://www.raeng.org.uk/news/publications/list/reports/Design_Engineering.pdf (accessed 21 November 2008).

28. Royal Academy of Engineering (2010) Engineering graduates for industry. Available athttp://www.raeng.org.uk/education/scet/pdf/Engineering_graduates_for_industry_report.pdf (accessed 17 June 2012).

29. Springer et al. (1999) Effects of small-group learning on undergraduates in science, mathematics, engineering and technology. A meta-analysis. Review of Educational Research 69 (1), 21–51. ,

30. Stump, G., Hilpert, J., Husman, J., Chung, W.-T., Kim, W. (2011) Collaborative learning in engineering students: gender and achievement. Journal of Engineering Education 100 (3) 475–497. ,

31. Tempelman, E. and Pilot, A. (2011) Strengthening the link between theory and practice in teaching design engineering: an empirical study on a new approach. International Journal of Technology and Design Education 21, 261–275. ,

32. UNESCO and Sustainable Development (2005). Available at http://unesdoc.unesco.org/images/0013/001393/139369e.pdf(accessed 18 August 2012).

33. Vygotsky, L.S. (1978) Mind in society: the development of higher psychological processes. London: Harvard University Press.

34. Willmot, P., Pond, K., Loddington, S. P. and Palermo, O. A. (2008) Perceptions of peer assessment in university teamwork. In International Conference on Engineering Education 27–31 July 2008. Pécs-Budapest, Hungary.

35. Zepke, N. and Leach, L. (2010) Improving student engagement: ten proposals for action. Active Learning in Higher Education 11 (3), 167–177.

36. Zhao, C.-M. and Kuh, G. (2004) Adding value: learning communities and student engagement. Research in Higher Education 45 (2), 115–138. ,

Chapter 10

MECHANICAL DESIGN AND KINEMATIC SIMULATION OF AUTOMATED ASSEMBLY SYSTEM FOR RELAY

Zhenghao Liu and Hong Chen

College of Mechatronics and Control Engineering, Shenzhen University, Shenzhen, China

ABSTRACT

By means of Solid Works, three-dimensional model of automated assembly system was established, and kinematic simulation based on Solid Works Motion of assembly process for relay was performed. The simulation results proved the feasibility of mechanical design. Eventually, the productivity was estimated based on simulation analysis. The mechanical design provided a solution with high reference value to practical design of automated assembly system for relay.

INTRODUCTION

In modern product manufacturing, automated assembly technology has been one of the key technologies [1] . The automated assembly technology is a comprehensive technology which combines the mechanical design, robotics and sensor and detection techniques [2] . As an important part for industrial production, the automated assembly can decide the total cost of production and productivity. And with the increasing functional demand for industrial products, more advanced automated assembly technology has access to meet the sophisticated demand [3] . Therefore, the automated assembly technology has become a hotspot of the current research. However, it exists some disadvantages of the automated assembly technology, such as inefficiency, high cost and complex mechanical structure [4] [5] .

There are seldom study work on mechanical design and kinematic simulation of the automated assembly technology in the current studies. An automated assembly system is studied in this paper. The system is designed for assembling a kind of relay by the machinery. Based on the mechanical structure of the assembly system, the whole working process of assembling the relay is achieved.

This paper is organized as follows. Principles of automated assembly system for relay are clarified in Section 2; the mechanical structure is illustrated as well. Section 3 focuses on the mechanical design of the Yoke, Bobbin and Base assembly part, including feed design, location design, clamping design and assembly design for the Yoke, Bobbin and Base. Kinematic simulation results and productivity estimation are shown in Section 4. Finally, conclusions are presented in Section 5.

PRINCIPLES OF AUTOMATED ASSEMBLY SYSTEM FOR RELAY

The main purpose of this project is to design a new practical automated assembly system for relay to meet work requirements of enterprises production. The working process of the assembly system is showing in Figure 1.

The procedure for each component involves eight steps and can be described as follows. At first, Bobbin is inserted into Yoke. The assembly of Yoke & Bobbin inserts into Base together in the next step. Thirdly, Hinge embeds in semi-manufactures. And Armature embeds in semi-manufactures in the fourth step. In the fifth step, SCC inserts into semi-manufactures. B-T and M-T insert into semi-manufactures in the sixth step. Then card covers semi-manufactures in the seventh step. Finally, case covers semi-manufactures. Since then, an intact relay is finishing assembling.

MECHANICAL DESIGN OF AUTOMATED ASSEMBLY SYSTEM FOR RELAY

Mainly Part of Automated Assembly System for Relay

To facilitate the mechanical design of the automated assembly system for relay, the working process of assembly system can be mainly divided into three parts, the Yoke, Bobbin and Base assembly part, the Hinge and Armature assembly part and the SCC, B-T and M-T assembly part, which can be shown in Figure 2.

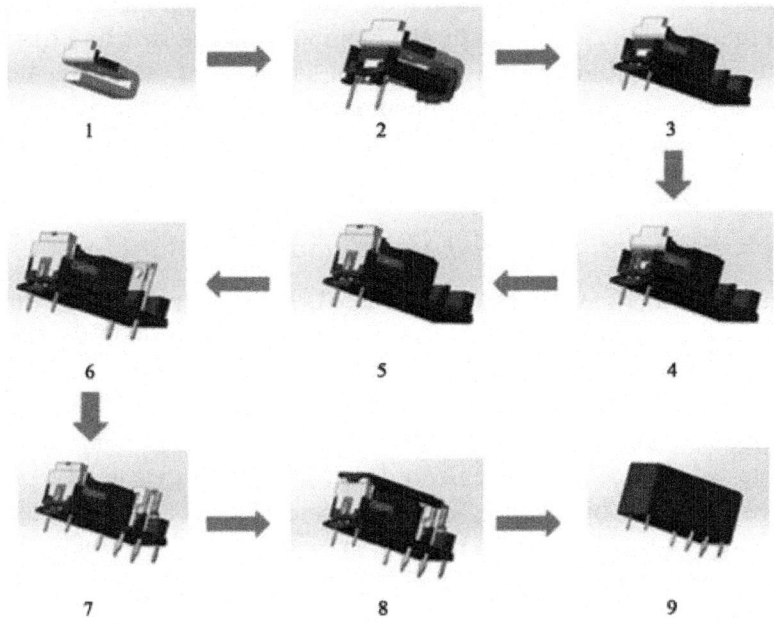

Figure 1: The working process of automated assembly system for relay.

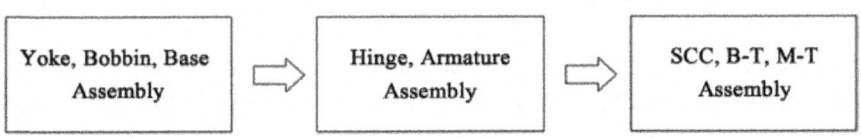

Figure 2: The mainly part of automated assembly system for relay.

In this paper, we choose the mechanical design of Yoke, Bobbin and Base assembly part to analysis.Figure 3 shows the overall mechanical design of Yoke, Bobbin and Base assembly part. This part consists of eight customized sub-structures. Each sub-structure's design aims at different shapes of components made up the relay, and necessary mechanical actions realized the assembly function.

Yoke, Bobbin and Base Assembly Part Design

1) Mechanical design of feed, location and clamping for Yoke

It's unnecessary to protect the surface and structure of Yoke in particular, so the combination of vibration tray and linear feeder is chosen to be the means for transporting Yoke. And Yokes in the linear slider stand closely in a row so that are inconvenient to capture by the manipulator. The structure of location for Yoke is designed to separate the Yoke in the head of the linear slider and the Yoke in the next, then the manipulator can clamp the Yoke easily. The Air TACHF Series Gripper [6] is effective and stable for clamping as a clamping component.

2) Mechanical design of feed and clamping for Bobbin

Compared with the feed of Yoke and Base, Bobbins surrounded by very thin copper wires get damage easily in transmission for assembly in the next step. A specialized pallet with 12×12 rooms for storing Bobbins is designed as a feeder. Because of the discontinuous way for feeding Bobbins by the pallet, a feed structure is customized to load the pallet filled with Bobbins and unload the empty pallet once automatically at least. This structure can take place the exchange for pallets by manual, as a result time is saved and the efficiency of the assembly system is improved. The clamping structure for Bobbin consists of two sets of step motor and a ball screw so that can complete the action of clamping for144 locations in the pallet.

3) Mechanical design of assembly for Yoke and Bobbin

As the accomplishment of feed and clamping of Yokes and Bobbins, the assembly structure will make them get together. In the procedure of assembly, the Bobbin is inserted into the Yoke and the Yoke as the base of the Yoke-Bobbin (the structure of assembly of Yoke & Bobbin). So the work table is designed to load the Yoke at first and then the Bobbin is inserted into it.

4) Mechanical design of feed for the Yoke-Bobbin

After the assembly of the Bobbin and the Yoke, the AirTAC HF Series Gripper is used for clamping the Yoke-Bobbin to the position for placing Bases, and the AirTAC STW Series Cylinder serves as the transmission actuator that transports the Yoke-Bobbin to the location of the Base. The clamping structure and transmission structure make up the feed structure for the Yoke-Bobbin.

5) Mechanical design of assembly for Base and Yoke-Bobbin

The assembly for the Base and the Yoke-Bobbin is the last step in this part of the assembly system. The feed for the Base is same as the Yoke that adopts the combination of vibration tray and linear feeder. Due to the different shape with Yokes, Bases are get separated by the simply dividing structure in the linear feeder without a customized structure for separating. And the location structure takes advantage of biaxial jaw to realize functions of clamping and location for the Base. In the procedure of assembly, the Yoke-Bobbin is inserted into the Base and the Base as the base of the first part assembly.

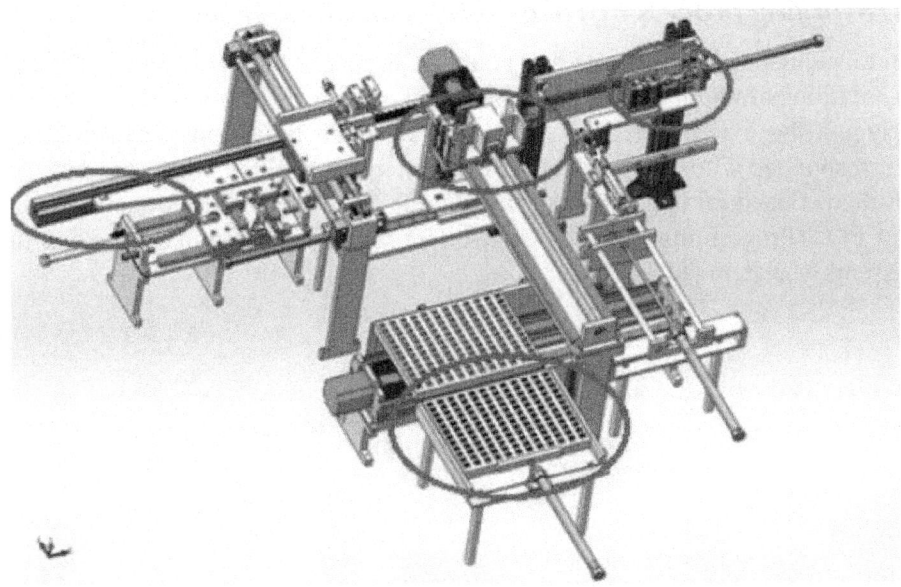

Figure 3: Diagram of Yoke, Bobbin and Base assembly part design.

The mechanical design of each sub-structure can be shown in the Figure 4. Each sub-structure can be described as follows. 1) The structure of location for Yoke; 2) The structure of clamping for Yoke; 3) The structure of feed for Bobbin; 4) The structure of clamping for Bobbin; 5) The work table of assembly for Yoke and Bobbin; 6) The structure of clamping for Yoke-Bobbin; 7) The structure of feed for Base; 8) The structure of assembly for Base and Yoke-Bobbin (The no. of each sub-structure is corresponded to the no. as shown in Figure 4).

KINEMATIC SIMULATION OF AUTOMATED ASSEMBLY SYSTEM FOR RELAY

After the three-dimensional model of the automated assembly system for relay is built. Kinematic simulation and analysis of the automated assembly system mainly focus on feasibility and validity of the mechanical design is implemented based on Solid Works Motion. In this section, we also choose the Yoke, Bobbin and Base assembly part to simulation and analysis.

Kinematic Simulation of Yoke, Bobbin and Base Assembly Part Design

1) Working process of Yoke, Bobbin and Base assembly part

In this Kinematic simulation and analysis, the event based motion is chosen as a solution paradigm [7] . In an event based motion, the motion of the assembly triggers the external action. And every event in the event based motion is corresponded with every mechanical action in the working process of assembly system. Based on the working process, the SFC (Sequential Function Chart) for PLC (Programmable Logic Controller) can be projected in related practical assembly system [8] .

1

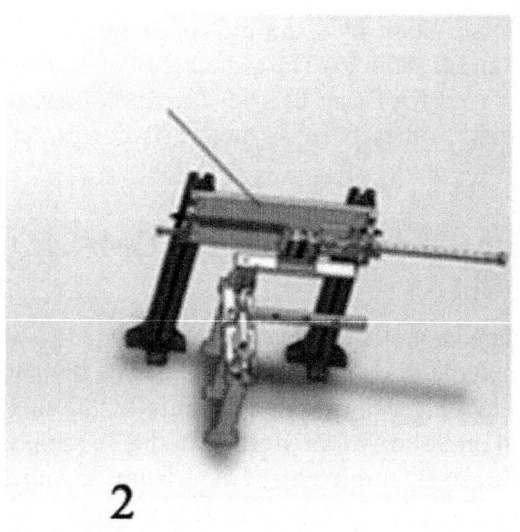

2

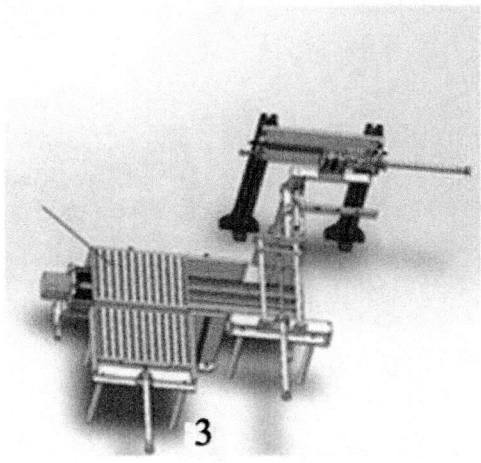

3

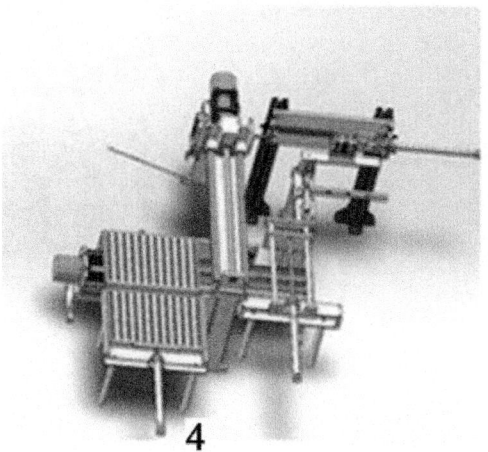

4

5

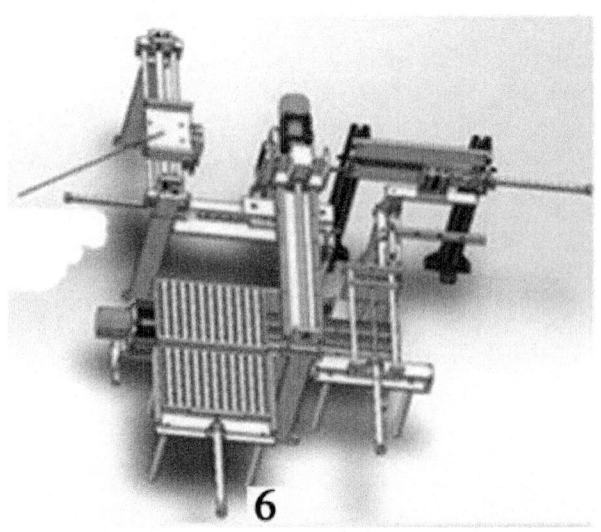

6

7

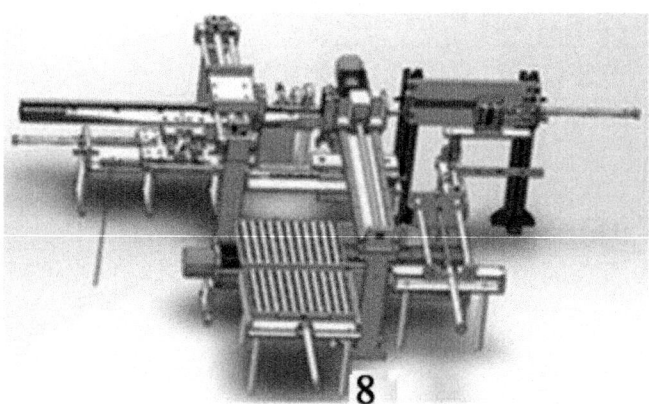

8

Figure 4: Diagram of each sub-structure design of Yoke, Bobbin and Base assembly part.

Analyzing the working process of the Yoke, Bobbin and Base assembly part, the SFC is exported easily which can be shown in the Figure 5. And it's clear to recognize the every task (event) in the event based motion of this kinematic simulation and analysis.

2) Motion analysis of Yoke, Bobbin and Base assembly part design

The steps for motion analysis of the Yoke, Bobbin and Base assembly part can be described as follows. Firstly, to get the assembly of this part applied to motion analysis, the overall assembly is disbanded into each sub-assembly layer by layer in the window of the model. Next, extra relationships of sub-assemblies are added in Solid Works and the immovable components are fixed. Then motion motors which include linear motor and rotation motor are added to the structure which is necessary to realize the mechanical action in the window of motion paradigm. Finally, every task for the whole part is added in the window of event and task at the base of the SFC for PLC of this assembly part.

Kinematic Simulation

Table 1 is parts of the task table of working process for Yoke, Bobbin and Base assembly part. This table is generated automatically by Solid Works after the motion analysis of the mechanical design.

According to the table, the function of each mechanical action which each task represents for and the sequence of every task has been shown. And it's obvious that how far mechanical structure has traveled and how much time has been consumed by each task in the working process of whole assembly part in the table.

The simulated animation of working process for Yoke, Bobbin and Base assembly part can get from the motion analysis. The animation displays every detail of working process for the assembly part totally. And the interference between the adjoining structures is easily to observe in the animation which shows the relationships among mechanical structures.

Parts of animation at different time of working process for the assembly part is shown in Figure 6, verifying that the Yoke, Bobbin and Base assembly part can work properly and the mechanical structure don't interfere with each other during the operation.

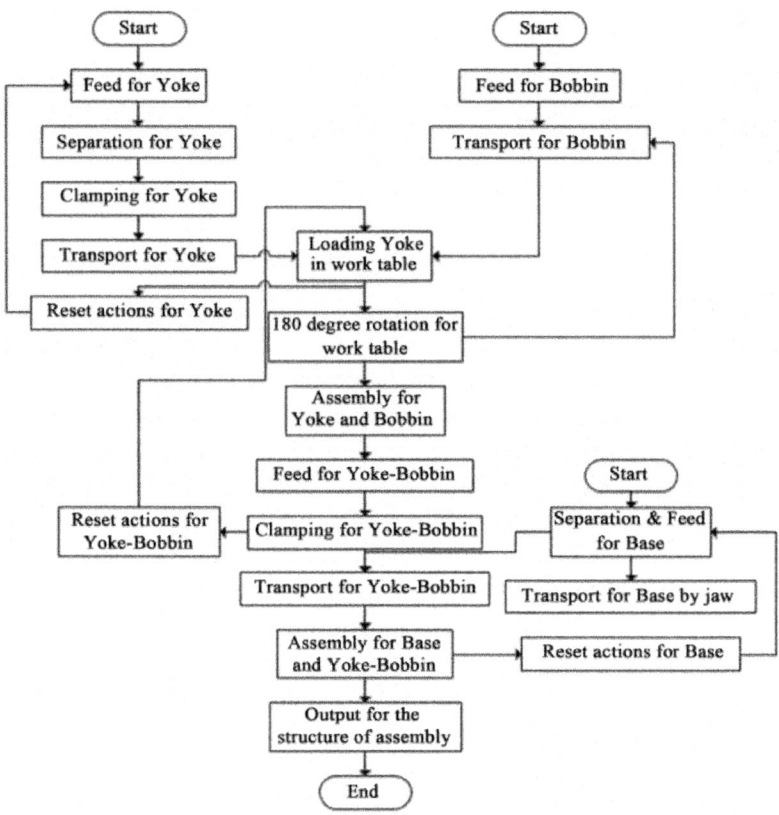

Figure 5: The SFC for PLC of Yoke, Bobbin and Base assembly part.

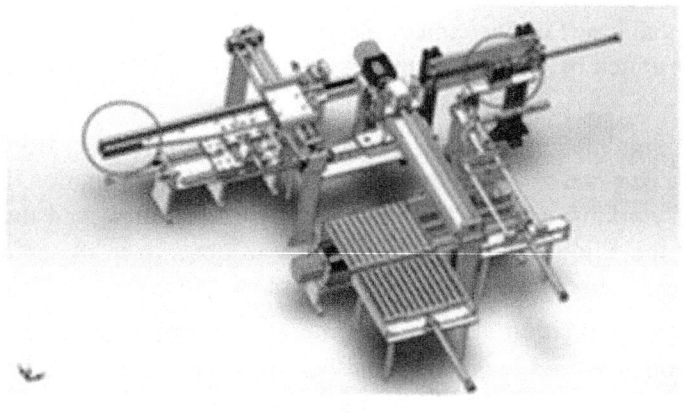

1

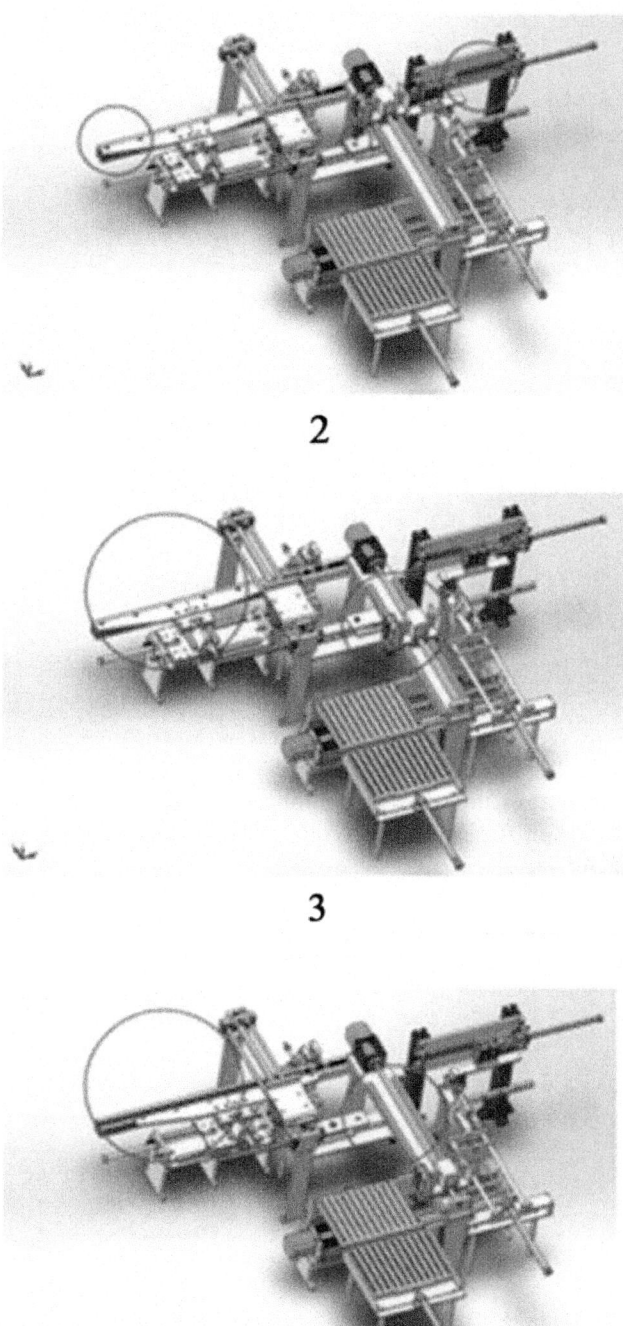

2

3

4

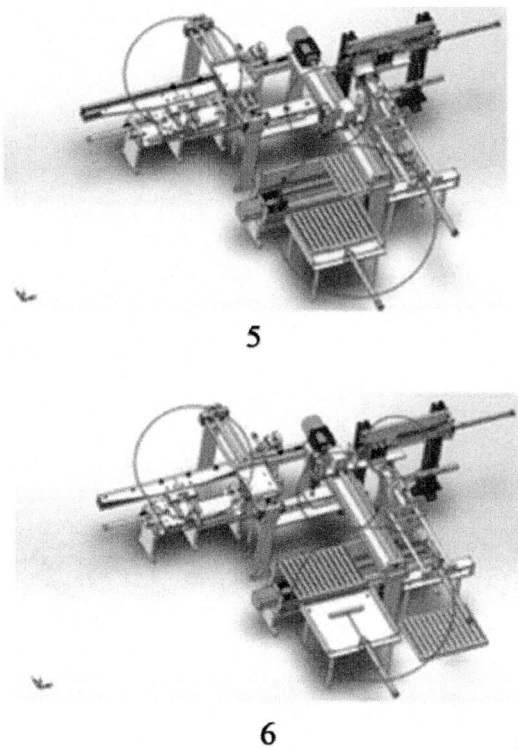

Figure 6: Screenshots of simulated animation at different time of the working process. 1. at 0 s; 2. at 5 s; 3. at 10 s; 4. at 20 s; 5. at 30 s; 6. at 40 s (Observe the change in the same cycle carefully).

Table 1: A part of task table of working process for Yoke, Bobbin and Base assembly part

Task	Trigger	Characteristic	Parameter	Time Start (s)	End (s)	Duration (s)
NO.169 Task	NO.168 Task	Base_3Z	36.31 (mm)	25.5	26	0.5
NO.170 Task	NO.169 Task	Location of Pallet	20 (mm)	26	26.5	0.5
NO.171 Task	NO.170 Task	Location of Base	−10 (mm)	26.5	27	0.5
		...				
NO.181 Task	NO.180 Task	Base_4Z	35.49 (mm)	26	28	0.5
NO.182 Task	NO.162 Task	Pallet for Yoke	−150 (mm)	26	28	2
NO.183 Task	NO.162 Task	Pallet for Yoke	−180 (deg)	26	28	2
		...				
NO.193 Task	NO.191 Task	Bobbin_5 Y	55.7 (mm)	31.5	32	0.5
NO.194 Task	NO.191 Task	Up and down clamping for Bobbin	−55.7 (mm)	31.5	32	0.5
NO.195 Task	NO.194 Task	Right and left clamping for Bobbin	−388.7 (mm)	32	34	2
		...				

Productivity Estimation for Yoke, Bobbin and Base Assembly Part

The pneumatic equipment usually works at the pressure of 0.4 MPa to 0.6 MPa [9] . Because of the low load for driving of every actuator, the time of most cylinders that achieve a single motion is close to others in practice. The productivity estimation for the Yoke, Bobbin and Base assembly part can be calculated as

$$(0.5+1) \times 2 \times 1.3 \div 2 = 1.95s \approx 2s \tag{1}$$

$$(0.5+1) \times 2 \times 1.3 \div 2.5 = 1.56s \approx 1.6s \tag{2}$$

where, 0.5 represents the simulation time of assembly for Yoke and Bobbin. And 1 represents the simulation time of assembly for Yoke & Bobbin and Base, both provided by the motion analysis of Solid Works. 2 stands for the assembly times of whole part assembly process for once. 1.3 is the coefficient of correction. Equation (1) and Equation (2) are corresponded to different amplification coefficients of simulation time in the whole motion analysis, which are 2 and 2.5 [10] .

To sum up, on the stable work condition of this assembly part, about 1.6 s to 2 s is consumed to finish the per assembly of Yoke, Bobbin and Base. The productivity is in line with the actual requirements of enterprise.

CONCLUSION

In this paper, mechanical design and kinematic simulation are researched. Principle and mechanical structures of the automated assembly system for relay are presented. The mechanical design of the Yoke, Bobbin and Base assembly part is performed especially, mechanical design of feed, location, clamping and assembly for the Yoke, Bobbin and Base are included. Kinematic simulation and analysis of the Yoke, Bobbin and Base assembly part focused on working process in practical assembly are performed based on Solid Works Motion. The simulation results verify the validity of the mechanical design of the Yoke, Bobbin and Base assembly part. This paper provides an effective and reliable mechanical design for the automated assembly system for relay.

ACKNOWLEDGEMENTS

The authors would like to thank the 2015 Teaching Research Project of Shenzhen University under the code 00002409260803 and Churod Electronics Co., LTD in Dongguan, Guangdong, especially the help and advice have provided by the mechanical engineers and electrical engineers.

REFERENCES

1. De Fazio, T.L., Edsall, A.C. and Gustavson, R.E. (1991) A Prototype of Feature-Based Design for Assembly. Springer, Berlin Heidelberg, 369-392. http://dx.doi.org/10.1007/BFb0014287

2. Wang, L., Keshavarzmanesh, S., Feng, H.-Y. and Buchal, R.O. (2009) Assembly Process Planning and Its Future in Collaborative Manufacturing: A Review. International Journal of Advanced Manufacturing Technology, 41, 132-144. http://dx.doi.org/10.1007/s00170-008-1458-9

3. Bi, Z., Wang, L. and Lang, S. (2007) Current Status of Reconfigurable Assembly Systems. International Journal of Manufacturing Research, 2, 303-328. http://dx.doi.org/10.1504/IJMR.2007.014727

4. Boer, C.R., Pedrazzoli, P., Sacco, M., Rinaldi, R., De Pascale, G. and Avai, A. (2001) Integrated Computer Aided Design for Assembly Systems. CIRP Annals, 50, 17-20. http://dx.doi.org/10.1016/S0007-8506(07)62061-7

5. Zhu, W., Wang, C., Wu, D., Tan, H. and Yang, F. (2009) Branch Aircraft's Digital Assembly Process Design and Simulation. IET International Communication Conference on Wireless Mobile and Computing, Shanghai, 7-9 December 2009, 413-416.

6. http://www.airtac.com/en/pro.aspx?c_kind=4&c_kind2=19&c_kind3=41&c_kind4=54/

7. http://www.solidworks.com/sw/products/simulation/motion-analysis.html

8. Yu, J., Wang, C., Yu, H. and Zhang, W. (2009) Generation of Optimized Assembly Sequences Based on Priority Rules Screening. Applied Mechanics and Materials, 16, 130-134. http://dx.doi.org/10.4028/www.scientific.net/AMM.16-19.130

9. Xu, L.D., Wang, C.G., Bi, Z.M. and Yu, J.P. (2012) Auto Assem: An Automated Assembly Planning System for Complex Products. IEEE Transactions on Industrial Informatics, 8, 1551-1561. http://dx.doi.org/10.1109/TII.2012.2188901

10. Jia, C.H., Liu, Y.B. and Xia, X.T. (2009) Research and Application of Digital Assembly Process Planning and Simulative Validation. International Conference on Mechatronics and Automation, Changchun, 9-12 August 2009, 2203-2207.http://ieeexplore.ieee.org/xpl/login.jsp?tp=&arnumber=5246737&tag=1&url=http%3A%2F%2F ieeexplore.ieee.org%2Fxpls%2Fabs_all.jsparnumber%3D5246737%26tag%3D1

Chapter 11

EVALUATION OF THE ULTIMATE CAPACITY OF FRICTION PILES

Wael N. Abd Elsamee

Faculty of Engineering, Sinai University, El Arish, Egypt

ABSTRACT

The precise prediction of maximum load carrying capacity of bored piles is a complex problem because the load is a function of a large number of factors. These factors include method of boring, method of concreting, quality of concrete, expertise of the construction staff, the ground conditions and the pile geometry. To ascertain the field performance and estimate load carrying capacities of piles, in-situ pile load tests are conducted. Due to practical and time constraints, it is not possible to load the pile up-to failure. In this study, field pile load test data is analyzed to estimate the ultimate load for friction piles. The analysis is based on three pile load test results. The tests are conducted at the site of The Cultural and Recreational Complex project in Port Said, Egypt. Three pile load tests are performed on bored piles of 900 mm diameter and 50 m length. Geotechnical investigations at the site are carried out to a maximum depth of 60 m. Ultimate capacities of piles are determined according to different methods including Egyptian Code of practice (2005), Tangent-tangent, Hansen (1963), Chin (1970), Ahmed and Pise (1997) and Decourt (1999). It was concluded that approximately 8% of the ultimate load is resisted by bearing at the base of the pile, and that up to 92% of the load is resisted by friction along the shaft. Based on a comparison of pile capacity predictions using different method, recommendations are made. A new method is proposed to calculate the ultimate capacity of the pile from pile load test data. The ultimate capacity of the bored piles predicted using the proposed method appears to be reliable and compares well to different available methods.

INTRODUCTION

Pile foundation is an important link in transferring the structural load to the bearing ground located at some depth below ground surface. The design of piles accounts for various parameters such as the nature of substrata, depth of ground water table, depth of the bearing stratum, and type and level of load to be supported. To ascertain the field performance and estimate the load carrying capacity, in-situ pile load tests are relied upon.

A simple method for calculating static shaft resistance of a pile driven into clay is presented by Mirza (1997) [1]. The method is based on correlations derived for marine clays between index properties and strengths. Applications of the method to half a dozen full scale pile load tests of high quality are described. Except for short piles in very stiff to hard clays, the predictions agree well with the field test measurements. The correlation presented allows an assessment of residual skin friction and indicates the importance of the liquidity index of the clay in static capacity calculations.

Dewaikar and Pallavi (2000) presented analysis of field pile load tests data to estimate the ultimate pile load. The analysis is based on forty pile load tests results collected from various infrastructure and building sites in Mumbai region of India. Collected data is analyzed using various graphical and semi-empirical methods available in literature [2].

Nabil (2001) studied the behavior of bored pile groups in cemented sands by a field testing program at a site in South Surra, Kuwait. The program consisted of axial load tests on single bored piles in tension and compression. Two groups of piles, each consisting of five piles were tested. The spacing between the piles in the groups was two and three-pile diameters. The calculated pile group efficiencies were 1.22 and 1.93 for a pile spacing of two and three-pile diameters, respectively. Since settlement usually controls the design of pile groups in sand, the group factor, defined as the ratio of the settlement of the group to the settlement of a single pile at comparable loads in the elastic range, was determined from test results [3].

Abdelrahman et al. (2003) suggested that axial pile loading tests on single pile may offer the justification of the pile design load. Codes for deep foundations design stipulate the acceptance criteria for piles tested in compression based on specified limits for pile settlement at specified load levels. The researchers examined the different methods used in interpreting pile load test results. Sixty-four continuous flight auger piles were tested using the maintained load test method and the results were analyzed using the different methods of interpretation [4].

Wehnert and Vermeer (2004) analyzed the load results of short large diameter bored pile tested in Germany. The results for total resistance as well as for base and shaft resistance are presented. The pile is assumed to be linear elastic. Different constitutive models for the subsoil such as elastic-plastic, Mohr-Coulmb, are used [5].

A new approach for the design of large diameter bored piles resting on cohesionless soils was suggested by Radwan et al. (2007) [6]. The approach is based on the results obtained from finite element analysis performed using data from thirty case histories of large diameter bored piles collected from several construction projects. Both unit end bearing and skin friction resistance are estimated taking the settlement criterion into account. Mohr-Coulomb constitutive model is used in the numerical model. Eventually, statistical study is conducted to evaluate the improvement, accuracy, and reliability of design using the new approach, compared with the prediction of the Egyptian Code (2005) [7].

Akbar et al. (2008) presented the experience gained from four pile load tests at a site in the North West Frontier Province of Pakistan. Geotechnical investigations at the site are carried out to a maximum depth of 60 m. The soil at the site is predominantly hard clays within the investigated depth with thin layers of gravels and boulders below 40 m depth. Four piles of diameters varying from 660 mm to 760 mm and length ranging between 20 m and 47.5 m were subjected to axial loads. Using the pile load test results, back calculations are carried out to estimate the appropriate values of pile design parameters [8].

A probabilistic model as a complementary mathematical base for the traditional deterministic approach to quantify the selection of a factor of safety for each term of the load equation of friction piles in clay is presented by Al Jairry (2009) [9].

From the above, the variation in the load estimates of available methods is too much. Thus, additional study on friction pile capacity is needed to be done. However, the objective of this study is to provide the results of pile tests and develop a formula for closer prediction of the pile capacity.

SOIL INVESTIGATION

There have not been many tests on the soil in Port Said in Egypt. The investigated site is the Cultural and Recreation Complex project located in the city of Port Said. The project is built on an area of approximately 50 × 70 m. A comprehensive geotechnical investigation was conducted. The investigation included seven borings. The general layout of site is shown in **Figure 1**.

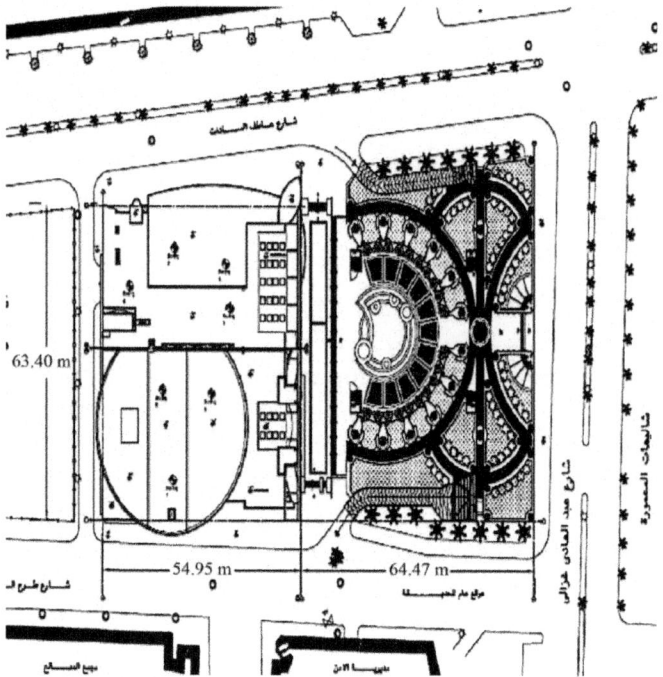

Figure 1: General layout of the site.

Soil Stratification

The soil profile in the investigated site is shown in **Figure 2**. The profile indicates that the following soil stratifications are encountered:

1.   From elevation 0.00 to –10.00 m calcareous sitysand with broken shells.
2.   From elevation –10.00 to –17.00 m soft silty-clay with interval of sand.
3.   From elevation –17.00 to –49.00 m soft silty-clay with traces of sand.
4.   From elevation –49.00 to –52.00 m calcareous sitysand.
5.   From elevation –52.00 to –60.00 m hard silty-clay with intervening calcareous silty-sand.

The ground water table has been found to be at 0.70 meter from the ground surface.

PREDICTION OF PILE LOAD CAPACITY USING EGYPTIAN CODE

Various field and laboratory tests are carried out during the geotechnical investigation for the evaluation of subsurface conditions and the pile design parameters at the project site. The code pile capacities are calculated using the provisions of the Egyptian code (2005) [7]. The pile diameter is taken as 900 mm and pile length is 50 m. Tables 1-3 summarize the soil properties as well as outlining the calculated pile resistance (shaft friction and end bearing). **Figure 3** shows the calculated ultimate capacity of the pile. Based on data from the figure, the ultimate pile capacity, Q_{ult} is obtained as 4622.81 kN/m². By applying a factor of safety, F.S. of 2, the allowable design pile capacity, Q_{all} is 2311.41 kN/m². The allowable bearing capacity of the pile adopted for the design is taken as 2300 kN/m².

PILE LOAD TESTS

Three pile load tests are performed on bored piles of 900 mm diameters and 50 m lengths. One of the piles is non-working pile test #1 and two are working piles tests #2 and #3. The nonworking pile test #1 is loaded to twice the working load of 230 ton while the working piles for tests #2 and #3 are loaded to 1.5 times the working load.

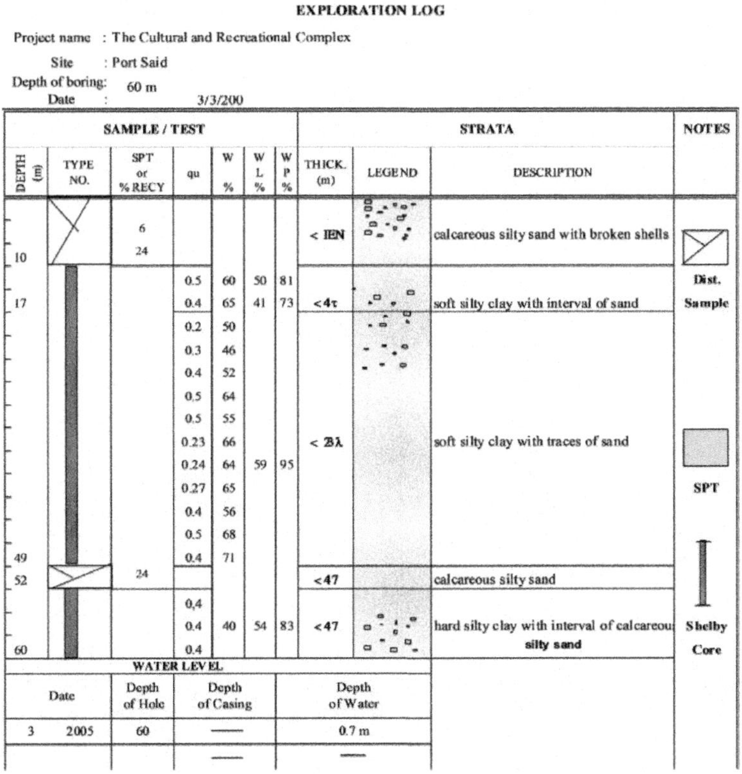

Figure 2: Soil profile of the investigated site.

Table 1: Calculated skin friction to be used in the design of pile according to the Egyptian Code [7].

Layer	Layer depth under the SBL [m]	Soil type	Av. SPT N value	Undrained cohesion Cu [kN/m²]	Depth [m]	SPT	Layer thickness [m]	Skin friction τ [kN/m²]	Friction pile load Q [KN]
1	0 - 5	CS-S	6		-	-	3	0	0.0
2	5 - 10	CS-S	24	-	2 - 7.5	20 - 30	5	75	1060.3
3	10 - 17	SS-C		20			7	20	395.8
4	17 - 49	HS-C		20	>7.5		32	20	1809.6
7	49 - 52	CS-S	>50		-	>50	2.1	100	593.8

Skin friction at settlement of 0.2 Sg = 0.9 cm, Qt = 3859.4; For Sg = 5%, D = 4.5 cm.

Table 2: Calculated end bearing resistance to be used in the design according to the Egyptian Code [7].

Point	Settlement [cm]		Bearing stress [KN/m²]	Pile area [m²]	End bearing pile load [KN]
O	0	0	0	0	0
A	0.2 Sg	1	500	0.64	318.09
B	0.3 Sg	1.35	700	0.64	445.32
C	Sg	4.5	1200	0.64	763.41

Table 3: Total pile load to be used in the design according to the Egyptian Code [7]

Point	End bearing pile load [KN]	Friction pile load Q KN]	Total pile resistance
O	0	0	0
A	318.09	3859.4	4177.49
B	445.32	3859.4	4304.72
C	763.41	3859.4	4622.81

Thus, the code ultimate capacity of pile = 3859.4 + 763.41 = 4622.81 kN/m².

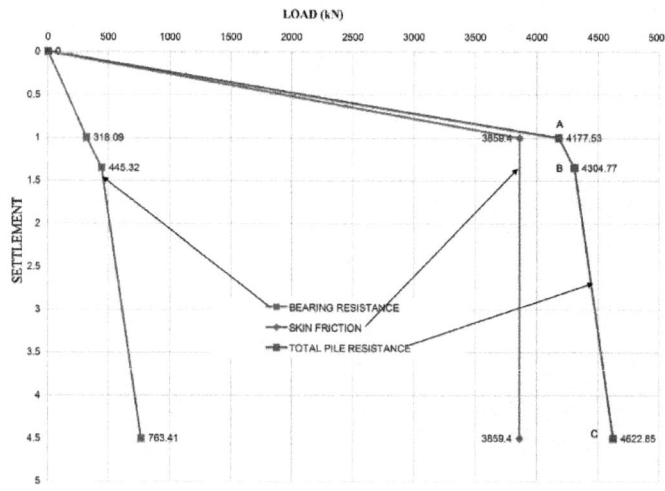

Figure 3: Shows the relationship between calculated capacity and settlement for the bored pile according the Egyptian Code.

Reaction System

The reaction system for the test piles was provided by a test head restrained by twelve ground anchors distributed around the pile as shown in the test setup in **Figure 4**.

Loading of Pile

The load was applied using three hydraulic jacks placed between the pile head and the anchored test head as shown in Figures 4 and 5. The loading cycles increment adopted for the test piles according Egyptian code.

Test Measurements

- Measurement of load The load was measured by calibrated load cells with digital readout device. Load cells were seated on top of spherical bearing plates placed above the hydraulic jacks. Also, the applied load was checked by recording the applied hydraulic pressure by a pressure gauge mounted on the pumping unit.

- Measurement of pile head settlement Settlement of the pile head is measured using three dial gauges of precision of 0.01 mm.

Test Results

1) General Observation during tests a) Settlement of pile did not reach 10% of its nominal diameter.

b) The test piles did not show any sign of geotechnical failure. This means that the test piles did not continue to settle or sink without increase in the applied load.

c) No section of the test piles failed structurally.

  The load-settlement relationships for pile load tests are shown in **Figure 6**.

d) Head Settlement is recorded in **Table 4**. It is noted that no sign of plunging is detected.

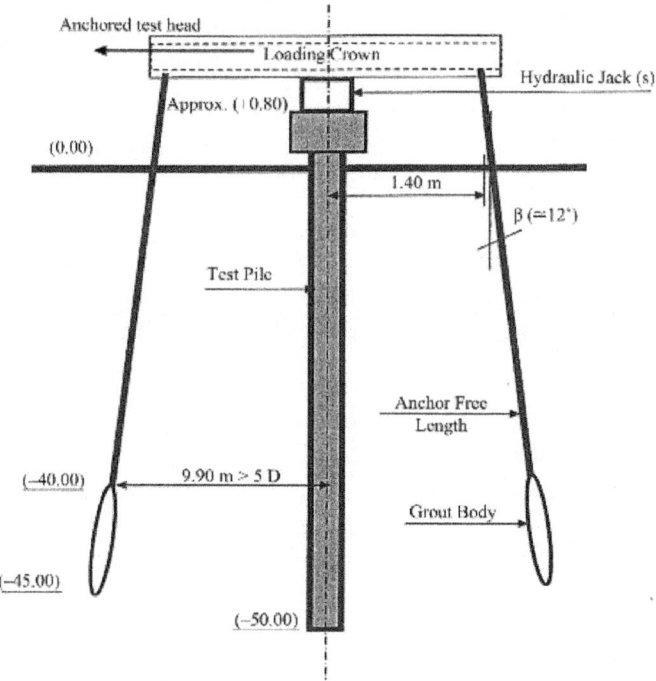

Figure 4: Test setup.

Figure 5: Pile loading setup.

ULTIMATE CAPACITY OF PILES

The ultimate capacities of the piles are determined from the load test results using different approaches.

Tangent—Tangent Method

Applying tangent—tangent method, a plot is made between load divided by cross sectional area of pile and the settlement on semi logarithmic scale as shown in **Figure 7** for working pile load test #2 [7].

Hansen Method (1963)

Applying Hansen Method the square root of each settlement value from field load test data divided by the corresponding load value is plotted against the settlement as shown in **Figure 8** for working pile load test #3. Estimation of the ultimate load by Hansen Method is given by the formula [10]:

$$Q_u = \left(2C_1C_2\right)^{1/2} \qquad (1)$$

where:

Q_u = ultimate load capacity.

C_1 = slope of the best fitting straight line.

C_2 = y-intercept of the straight line.

Chin's Method (1970)

Applying Chin's method, a plot is made between settlement divided by corresponding load and the settlement as shown in **Figure 9** for non-working test pile #1. The inverse slope of the straight line gives the ultimate load as proposed by Chin [11].

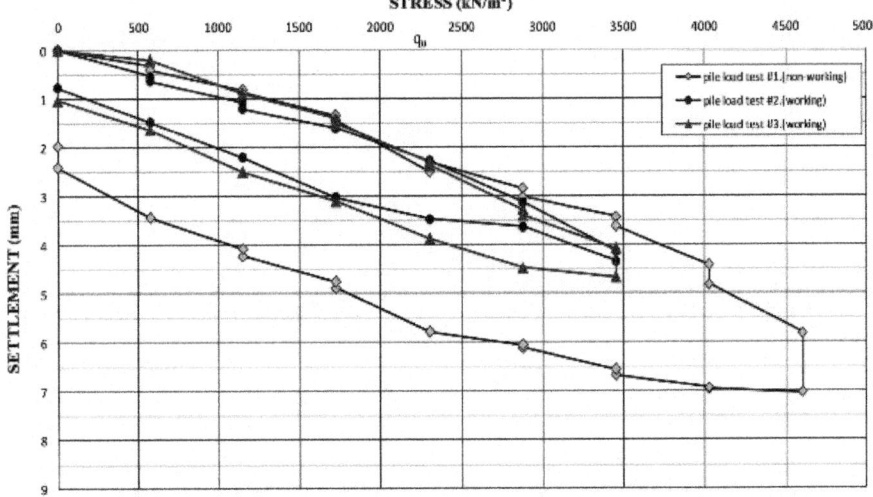

Figure 6: Load-settlement relationship for non-working pile load test #1.

Table 4: Recorded head settlement

Test No.	Pile #1 non-working	Pile #2 working	Pile #3 working
Settlement at 230 tons (anticipated working load)	2.27 mm	2.29 mm	3.40 mm
Settlement at 345 tons (150% of the working load)	3.62 mm	4.33 mm	3.87 mm
Settlement at 460 tons (200% of the working load)	7.03 mm	-	-
Residual (parameter) settlement	1.97 mm	0.77 mm	1.03 mm

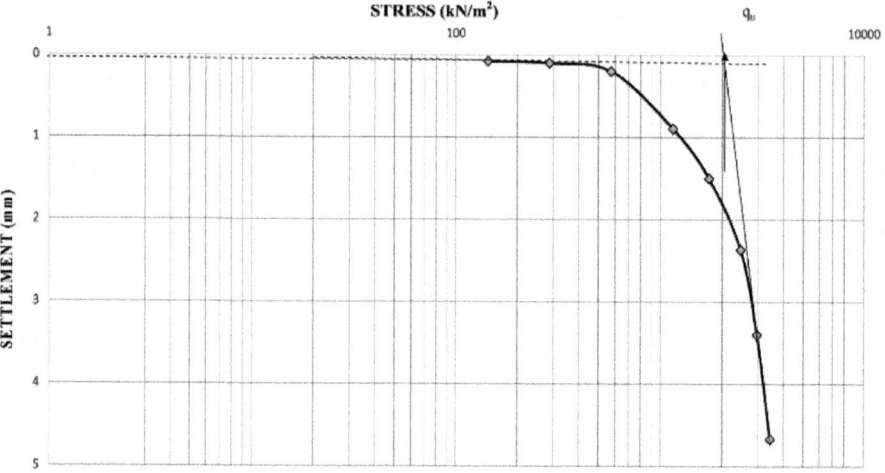

Figure 7: Ultimate pile capacities by tangent—tangent method for working pile load test #2.

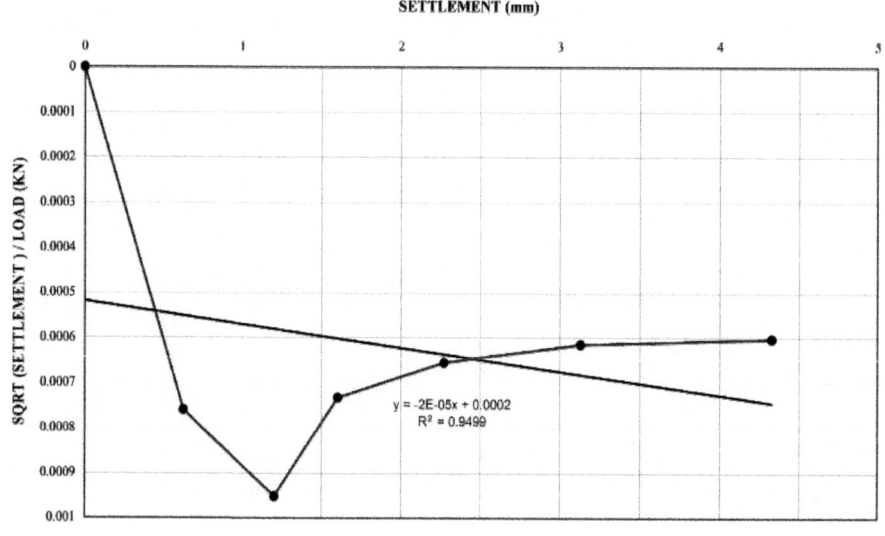

Figure 8: Ultimate pile capacities by Hansen method for working pile load test #3.

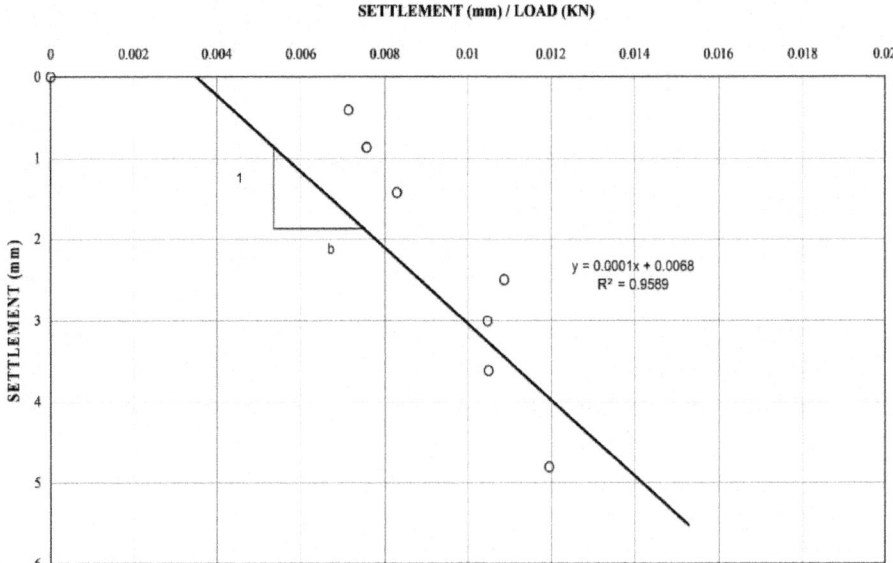

Figure 9: Ultimate pile capacity by Chin method for non-working test pile #1.

Ahmad and Pise (1997)

Ahmad and Pise (1997) proposed a reduction factor to Chin's extrapolated value of the ultimate capacity. In the settlement/load vs. settlement plot, it was observed that, generally two straight lines could be drawn through these points. As shown in **Figure 10** for non-working test pile #1, the ratio of settlement ΔS (settlement between the point of intersection of two straight lines and that corresponding to final test load) to S (total settlement) is taken to be the reduction factor (RF) for that set of test data [12]. However, reduction factor (RF) is given by the following:

$$RF = \frac{\Delta S}{S}$$

(2)

where:

RF = Reduction factor.

Q_{mod} = Modified Chin's value of ultimate capacity.

Q_{ch} = Chin's value of ultimate capacity.

Decourt's Extrapolation (1999)

Applying Decourt's Extrapolation by dividing each load by its corresponding settlement and ploting the resulting values against the applied load. A linear regression over the apparent line (last three points) determines a line. Decourt identified the ultimate load as the intersection of this line with load axis as shown in Figures 11 for working test pile #3 [13].

PROPOSED METHOD FOR DETERMINATION OF ULTIMATE PILE CAPACITY FROM LOAD TEST

The load vs settlement behavior of the pile is extrapolated using an empirical method. The estimation of ultimate load consists of two steps as given below:

1) Plotting load settlement curve from field load test data as shown in Figures 12-14.

2) The ultimate pile capacity is given by the empirical formula:

$$Q_u = \left[\frac{1}{0.445my} \right]$$

(3)

where:

Q_u = ultimate load capacity (kN).

m = slope of the trend straight line.

y = y-intercept of the straight line (as a value without sign).

COMPARISON BETWEEN DIFFERENT METHODS FOR DETERMINATION OF ULTIMATE PILE CAPACITY

The calculation of the ultimate capacity of piles and the corresponding factors of safety using the above mention methods are summarized in **Table 5**.

The ultimate loads obtained by various methods from the pile load test results are shown in **Figure 15**.

LOAD CARRIED BY END BEARING AND FRICTION ALONG SHAFT

From **Table 6** the values of the ultimate pile capacity were taken to evaluate the percentage of friction and end bearing capacity from **Figure 3**. Based on the above findings, it was found that the percentage of load carried by friction along the pile shaft and the end bearing are shown in the following **Table 6**.

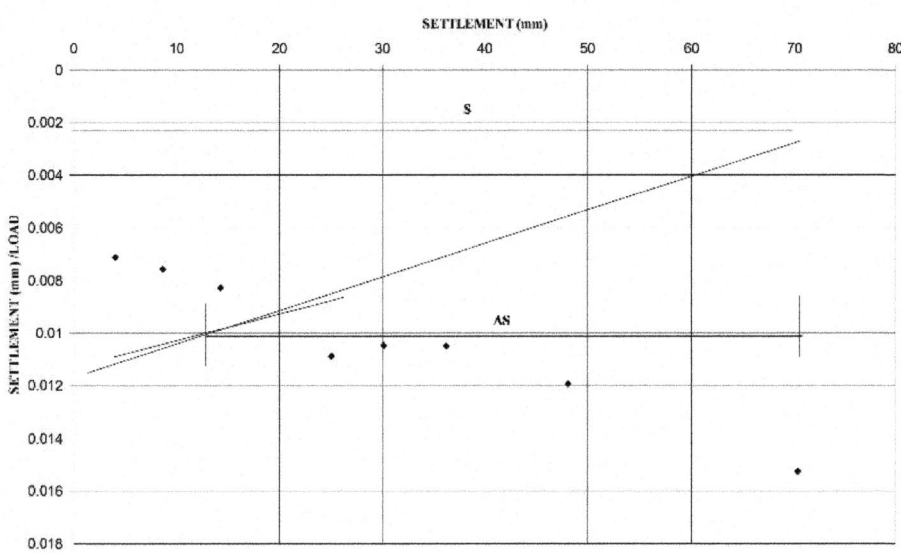

Figure 10: Ultimate pile capacity by Ahmad and Pise method for non-working test pile #1.

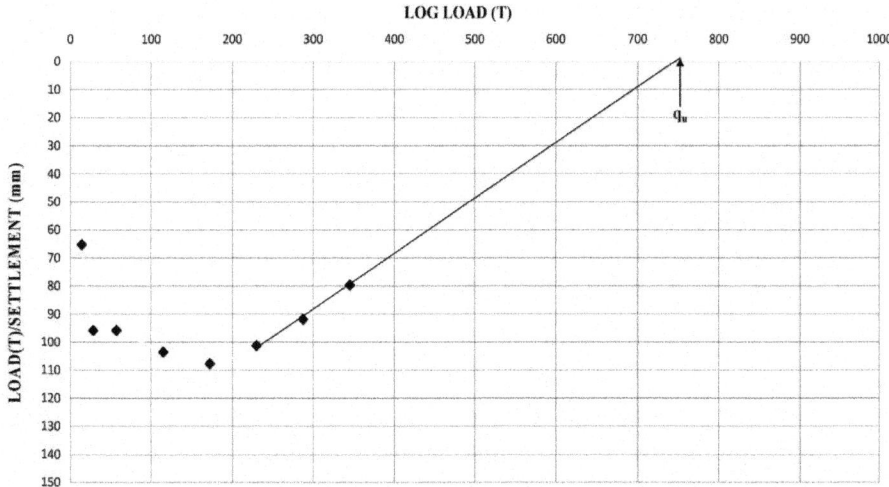

Figure 11: Ultimate pile capacities by Decourt's extrapolation method for working test pile #3.

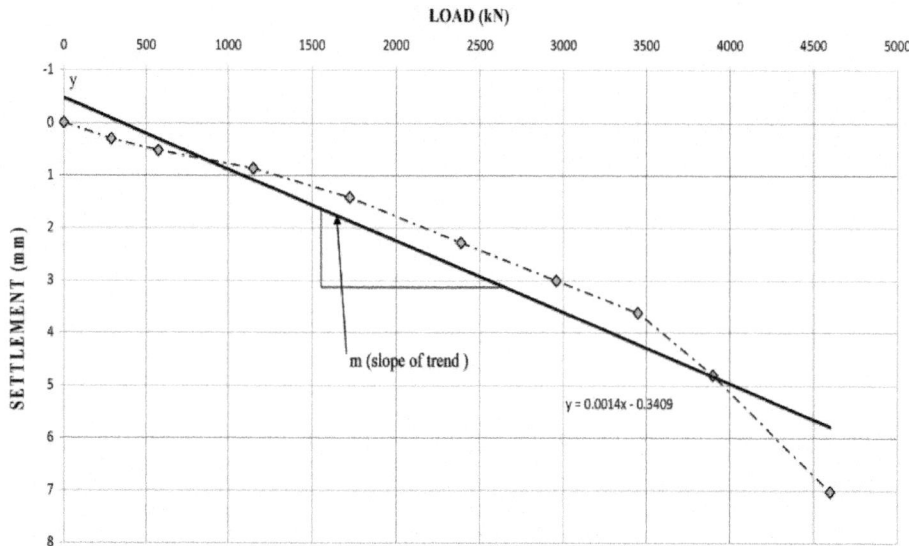

Figure 12: Ultimate pile capacity using proposed method for non-working test pile #1.

CONCLUSIONS

From the testing program and comparable study conducted, the following conclusions are arrived at:

1. The percentage of friction load carried by the shaft is approximately 85% to 90% and the percentage of load carried by the end bearing is 15% to 10%.

2. Hansen (1963) method gives higher values of ultimate capacity carried by the pile than the other methods.

3. A new proposed method to calculate the ultimate capacity of pile from pile load test is presented.

4. The proposed method for determining the ultimate capacity of friction piles appears to give results that are in good agreement with the analytical predictions.

5. The proposed method is good to apply, easier, quicker, more reliable, does not give max or min numbers as compared to some others.

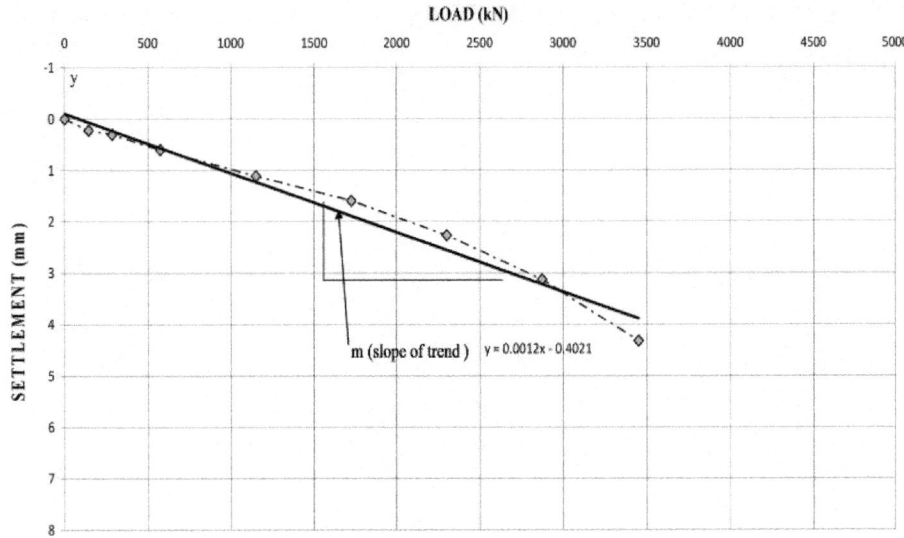

Figure 13: Ultimate pile capacity piles using proposed method for working test pile pile #2.

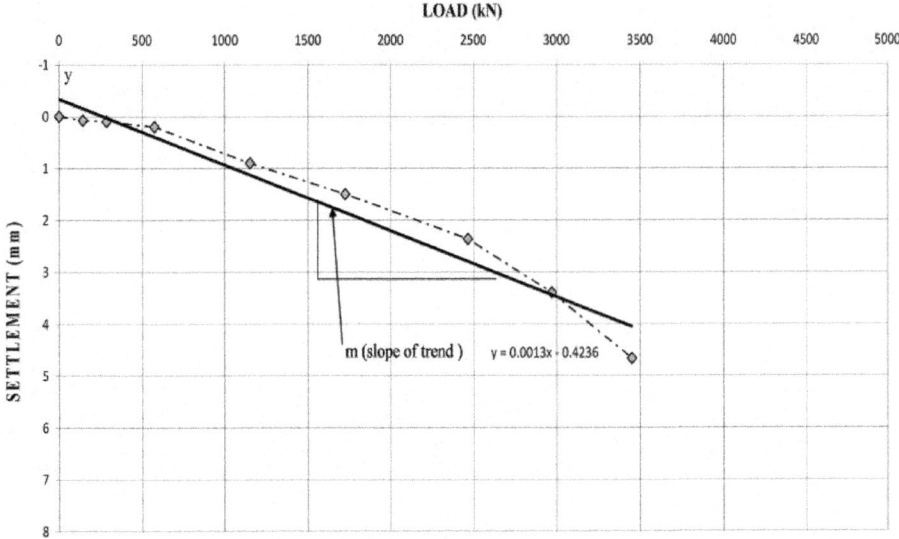

Figure 14: Ultimate pile capacity pile using proposed method for working test pile #3.

Table 5: Ultimate capacity and factor of safety (F.S.) of pile using different methods.

Test No.	Pile #1 non working		Pile #2 working		Pile #3 working	
Method	Q_{ult} (kN)	F.S.	Q_{ult} (kN)	F.S.	Q_{ult} (kN)	F.S.
Tangent	5600.00	2.43	5300.00	2.30	4400.00	2.00
Hansen (1963)	9128.71	3.97	5000.00	2.17	3227.49	1.40
Chin (1970)	8333.33	3.62	5555.56	2.14	4166.67	1.81
Ahmed & Pise (1997)	6641.66	2.88	4381.58	1.91	3319.06	1.44
Decourt (1999)	6990.00	3.03	7300.00	3.17	5750.00	2.50
Present study	4720.99	2.05	4658.36	2.03	4080.49	1.77

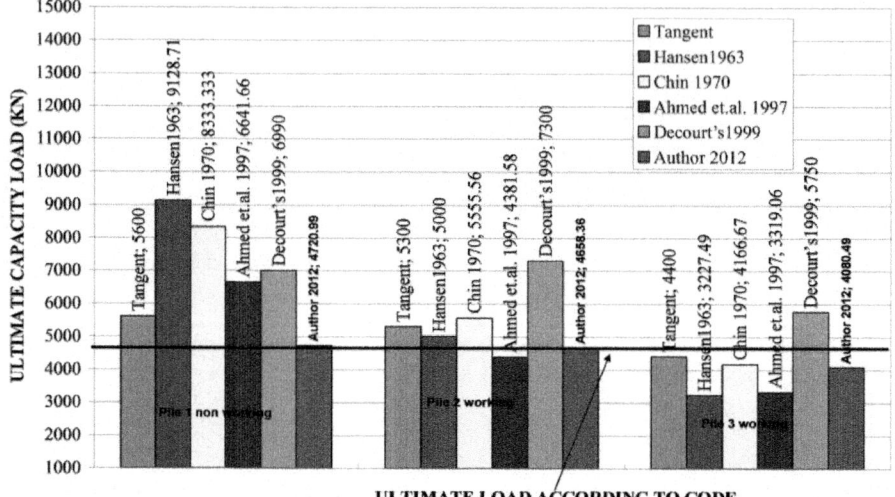

Figure 15: Comparison of ultimate pile loads using different methods.

Table 6: Percentage of ultimate load carried by end bearing and friction.

Test pile No.	Pile #1 non-working		Pile #2 working		Pile #3 working	
Method	Skin friction %	End bearing %	Skin friction %	End bearing %	Skin friction %	End bearing %
Code load	90.8	9.20	92.5	7.50	92.2	7.80
Tangent	84.3	15.7	83.9	16.1	87.2	12.8
Hansen (1963)	88.1	11.9	90.8	9.20	88.1	11.9
Chin (1970)	86.8	13.2	83.9	16.1	83.7	16.3
Ahmed and Pise (1997)	85.0	15.0	83.5	16.5	86.0	14.0
Decourt (1999)	85.4	14.60	85.70	14.30	84.10	15.90
Present study	91.2	8.80	92.3	7.70	90.6	9.4

ACKNOWLEDGEMENTS

The author would like to acknowledge the Fetih Construction Company and Pauer-Egypt Company for their valuable assistance.

REFERENCES

1. U. A. A. Mirza, "Pile Skin Friction in Clays," International Journal of Offshore and Polar Engineering, Vol. 7, No. 1, 1997, pp. 538-540.

2. D. M. Dewaikar and M. J. Pallavi, "Analysis of Pile Load Tests Data," Journal of Southeast Asian Geotechnical Society, Vol. 6, No. 4, 2000, pp. 27-39.

3. F. I. Nabil, "Axial Load Tests on Bored Piles and Pile Groups in Cemented Sands," Journal of Geotechnical and Geoenvironmental Engineering, Vol. 127, No. 9, 2001, pp. 766-733.doi:10.1061/(ASCE)1090-0241(2001)127:9(766)

4. G. E. Abdelrahman, E. M. Shaarawi and K. S. Abouzaid, "Interpretation of Axial Pile Load Test Results for Continuous Flight Auger Piles," Emerging Technologies in Structural Engineering, Proceedings of the 9th Arab Structural Engineering Conference, Abu Dhabi, 29 November- 1 December 2003, pp. 791-802.

5. M. Wehnert and P. A. Vermeer, "Numerical Analysis of Load Test on Bored Piles," Proceedings of the Ninth International Symposium on "Numerical Models in Geomechanics", Ottawa, 25-27 August 2004, pp. 1-6.

6. A. M. Radwan, A. H. Abdel-rahman, M. Rabie and M. F. Awad-Allah, "New Suggested Approach for Design of Large Diameter Bored Piles Based on Finite Element Analysis," Twelfth International Colloquium on Structural and Geotechnical Engineering (12th ICSGE), 10-12 December 2007, Cairo, pp. 340-357.

7. Egyptian Code, "Soil Mechanics and Foundation," Organization, Cairo, 2005.

8. A. Akbar, S. Khilji, S. B. Khan, M. S. Qureshi and M. Sattar, "Shaft Friction of Bored Piles in Hard Clay," Pakistan Journal of Engineering and Applied Science, Vol. 3, 2008, pp. 54-60.

9. H. H. Al Jairry, "Exact Probability Equation for Friction Piles in Clay," Iraqi Journal of Civil Engineering, Vol. 6, No. 1, 2009, pp. 791-802.

10. J. B. Hansen, "Discussion on Hyperbolic Stress-Strain Response, Cohesive Soils," Journal for Soil Mechanics and Foundation Engineering, Vol. 89, 1963, pp. 241- 242.

11. F. K. Chin, "Estimation of the Ultimate Load of Piles from Tests Not Carried to Failure," Proceedings of Second Southeast Asian Conference on Soil Engineering, Singapore City, 11-15 June 1970, pp. 81-92.

12. F. Ahmed and P. J. Pise, "Pile Load Test Data-Interpretation & Correlation Study," Indian Geotechnical Conference, Vadodara, 17-20 December 1997, pp. 443-446.

13. L. Decourt, "Behavior of Foundations under Working Load Conditions," Proceedings of the 11th Pan-American Conference on Soil Mechanics and Geotechnical Engineering, Foz DoIguassu, August 1999, Vol. 4, pp. 453-488.

Chapter 12

ATTITUDE CONTROL OF AN AXI-SYMMETRIC RIGID BODY USING TWO CONTROLS WITHOUT ANGULAR VELOCITY MEASUREMENTS PAPER

Tawfik El-Sayed Tawfik

Department of Mathematics, Faculty of Science, Mansoura University, Mansoura, Egypt

ABSTRACT

This paper considers the problem of controlling the rotational motion of an axi-symmetric rigid body using two independent control torques without angular velocity measurements. The control law which stabilizes asymptotically this motion is obtained only in terms of the orientation parameters. Global asymptotic stability is shown by applying LaSalle invariance principal. Numerical simulation is introduced

INTRODUCTION

A rigid body in general (non-symmetric) is controlled with three independent controls without angular velocity measurements [1-3]. If one of the controls is failure, the rigid body is not controllable. Thus the attitude control of a rigid body motion using two controls is an important control problem.

The angular velocity along the symmetric axis of the rigid body is fixed to its initial value. In this case, two control torques are used to stabilize asymptotically the rotational motion about the symmetric axis. Moreover, the orientation of the symmetric axis is described using stereographic coordinates form direction cosines [4].

Many authors have discussed the attitude control of a rigid body motion using two controls that depend in terms of the angular velocities of the rigid body and the orientation parameters. The stabilization of a zero total angular momentum satellite using two reaction wheels has been shown in [5,6]. Two controls which stabilize asymptotically a rigid body motion using matching

condition are obtained in terms of the angular velocities of the rigid body
[7]. Two controls which stabilize asymptotically an axi-symmetric rigid
spacecraft are obtained in terms of the angular velocities of the rigid body
and the orientation parameters [8-11]. The angular velocity measurement is
noisy. It contains high frequency and random fluctuations. In this paper, two
control torques which stabilize asymptotically the rotational motion of an axi-
symmetric rigid body are obtained only in terms of the orientation parameters.

The present paper is organized as follows: Section 2 presents dynamic
and kinematic equations of an axisymmetric rigid body with two control
torques. Section 3 is devoted to obtain the two control torques which stabilize
asymptotically the rotational motion of an axi-symmetric rigid body in terms of
the orientation parameters. The asymptotic stability of this motion is proved by
applying LaSalle invariance principal. Section 4 contains numerical simulation
to illustrate the theoretical results of the paper.

DYNAMICS AND KINEMATICS

Consider the rotational dynamics of an axi-symmetric rigid body controlled
by two independent control torques. Two reference frames are introduced. The
first $\hat{n} = (\hat{n}_1, \hat{n}_2, \hat{n}_3)$, is an inertial reference frame, and the second $\hat{b} = (\hat{b}_1, \hat{b}_2, \hat{b}_3)$ is
a body-fixed reference frame and coincident with the principal axes of inertia
of the body. The unit vector $\hat{b}_3$ lies along the axis of symmetry. Two control
torques u_1 and u_2 are applied along the unit vectors $\hat{b}_1$ and $\hat{b}_2$, respectively
(Figure 1). Let A_i and $\omega_i (i = 1, 2, 3)$ be the principal moments of inertia of the
rigid body and the components of the angular velocity of the body referred to
the $\hat{b}$ frame, respectively. The dynamic equations take the form:

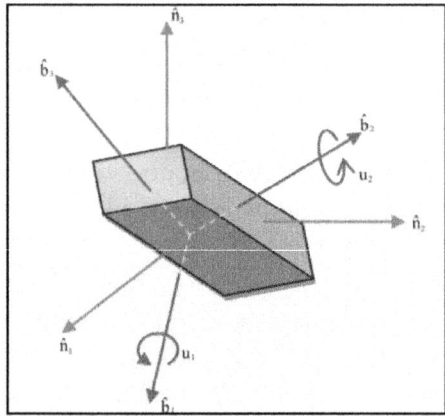

Figure 1: Axi-symmetric rigid body with two controls.

$$A_1\dot{\omega}_1 = (A_2 - A_3)\omega_2\omega_3 + u_1,$$
$$A_2\dot{\omega}_2 = (A_3 - A_1)\omega_3\omega_1 + u_2,$$
$$A_3\dot{\omega}_3 = (A_1 - A_2)\omega_1\omega_2.$$

$$(2.1)$$

Since $A_1 = A_2 = A$, , if we let the initial condition $\omega_3(0) = \omega_{30}$, $\dot{\omega}_3$ will remain constant throughout the maneuver. Equations (2.1) can rewrite as:

$$A\dot{\omega}_1 = (A - A_3)\omega_{30}\omega_2 + u_1,$$
$$A\dot{\omega}_2 = (A_3 - A)\omega_{30}\omega_1 + u_2.$$

$$(2.2)$$

The orientation of an axi-symmetric rigid body is described by using stereographic coordinates form direction cosines [4]. Two orientation parameters W_1 and W_2 can be used to describe the position of the $\hat{n}_3$ inertial axis in the body fixed $\hat{b}$ frame. These parameters satisfy the differential equation:

$$\dot{W}_1 = \omega_3 W_2 + \omega_2 W_1 W_2 + \frac{\omega_1}{2}(1 + W_1^2 - W_2^2),$$
$$\dot{W}_2 = -\omega_3 W_1 + \omega_1 W_1 W_2 + \frac{\omega_2}{2}(1 + W_2^2 - W_1^2).$$

$$(2.3)$$

equations (2.2) and (2.3) can be written in a vector form as:

$$A\dot{\omega} = (A - A_3)S(\omega_{30})\omega + u,$$

$$(2.4)$$

$$\dot{W} = S(\omega_{30})W + F(W)\omega$$

$$(2.5)$$

where

$$\omega = [\omega_1 \quad \omega_2]^T, W = [W_1 \quad W_2]^T, u = [u_1 \quad u_2]^T,$$ F (W) is the 2×2 symmetric matrix

$$F(W) = \frac{1}{2}((1 - W^T W)I + 2WW^T)$$

$$(2.6)$$

and $S(\omega_{30})$ is the 2 2 ´ skew-symmetric matrix

$$S(\omega_{30}) = \begin{bmatrix} 0 & \omega_{30} \\ -\omega_{30} & 0 \end{bmatrix}.$$

$$(2.7)$$

Equations (2.4) and (2.5) can be used to solve the problem of controlling the rotational motion of an axisymmetric rigid body, using Liapunov function technique.

The main objective is to determine the control law in terms of the orientation parameters that will derive u w and to zero. To derive this control law, we introduce the new parameters

$$\hat{W} = \begin{bmatrix} \hat{W}_1 & \hat{W}_2 \end{bmatrix}^T$$

which estimate the orientation parameters

$$W = \begin{bmatrix} W_1 & W_2 \end{bmatrix}^T,$$

respectively. Also we suppose that the orientation parameters and their estimates satisfy the following auxiliary system of differential Equation:

$$\dot{\hat{W}} = W - \hat{W} + S(\omega_{30})W.$$

(2.8)

Using the kinematic Equation (2.5) the auxiliary system (2.8) can be written in the form:

$$\dot{\xi} = -\xi + F(W)\omega$$

(2.9)

where

$$\xi = W - \hat{W}.$$

(2.10)

STABILIZATION PROBLEM

The main object of this section is to determine the control law u which stabilizes asymptotically the system (2.4), (2.5), (2.9). This control law depends upon the orientation parameters only.

Theorem. The control law

$$u = -k\left(W^T + \xi^T\right)F(W)$$

(3.1)

Where k>0 stabilizes asymptotically the system (2.4), (2.5), (2.9).

Proof

Assume that, the Liapunov function in the form

$$2\Phi = A\omega^T\omega + k\left(W^T W + \xi^T \xi\right).$$

(3.2)

This function is a positive definite with respect to stabilize variables w , W , and x . The time derivative of the Liapunov function (3.2) using (2.4), (2.5), (2.9) and the control law (3.1) takes the form

$$\frac{d\Phi}{dt} = A\omega^T\dot{\omega} + k\left(W^T\dot{W} + \xi^T\dot{\xi}\right)$$

$$= \omega^T\left(u + k\left(W^T + \xi^T\right)F(W)\right) - k\xi^T\xi$$

$$= -k\xi^T\xi \le 0.$$

$$(3.3)$$

The time derivative of the Liapunov function is a negative semi-definite function (constant sign function). Thus, under the control law (3.1), the system is stable.

Now, we will prove the asymptotic stability of this system using LaSalle Invariance Principle [12]. Define $\bar{\Omega}$ as the largest invariant set in

$$\Omega = \left\{(\omega, W, \xi) : \dot{\Phi} = 0\right\} = \left\{(\omega, W, \xi) : \xi = 0\right\}.$$

On $\bar{\Omega}$ we have that $\dot{\xi} = F(W)\omega = 0$ & from (2.9). This implies that w = 0 on $\bar{\Omega}$. Since $\bar{\Omega}$ is invariant, $\dot{\omega} = 0$ in turn implies

$$-kW^T F(W) = 0$$

(from (2.4) and (3.1)). This implies that W = 0 on $\bar{\Omega}$. Therefore

$$\bar{\Omega} = \left\{(\omega, W, \xi) : \omega = W = \xi = 0\right\}.$$

NUMERICAL SIMULATION

This section shows the effect of the value of the control constant k in control purposes. The Program used in this numerical approach is MAPLE. We choose the inertial moments of an axi-symmetric rigid body, the initial angular velocities of the rigid body, the initial orientation parameters and the initial error attitude parameters as follows:

$$A_1 = A_2 = 15, A_3 = 20 \text{ kg} \cdot \text{m}^2,$$

$$\omega_1(0) = 0.2, \omega_2(0) = 0.2, \omega_3(0) = 0.1 \text{ rad/s},$$

$$W_1(0) = 0, W_2(0) = 0,$$

$$\xi_1(0) = 0, \xi_2(0) = 0.$$

$$(4.1)$$

Figures 2(a)-(d) show the time response of the body angular velocities, the orientation parameters, the error of the orientation parameters and the control torques, respectively for the control constant k = 2 .

Figures 3(a)-(d) show the time response of the body angular velocities, the orientation parameters, the error of the orientation parameters and the control torques, respectively for the control constant k = 20

Figures 4(a)-(d) show the time response of the body angular velocities, the orientation parameters, the error of the orientation parameters and the control torques, respectively for the control constant k = 40 .

Based on the above numerical simulation study we conclude that, increasing the value of the control constant k has the effect of decreasing the convergence behavior of the system. This result is different from the result when the control torques are obtained in terms of the angular velocities of the rigid body and the orientation parameters [11].

CONCLUSION

The angular velocity measurement contains high frequency and noises or random fluctuations. Two control a torques (3.1) which stabilize asymptotically the rotational motion of an axi-symmetric rigid body are obtained in terms of the orientation parameters without angular velocity measurements. Global asymptotic stability is shown by applying LaSalle invariance principal.

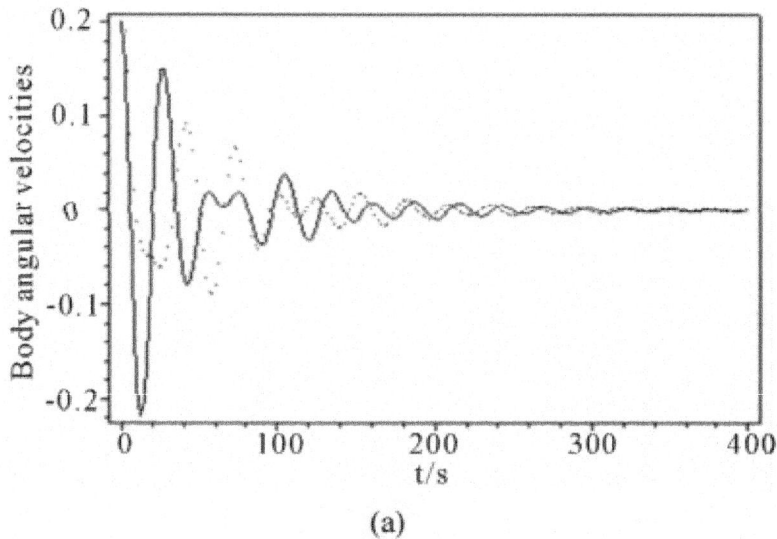

(a)

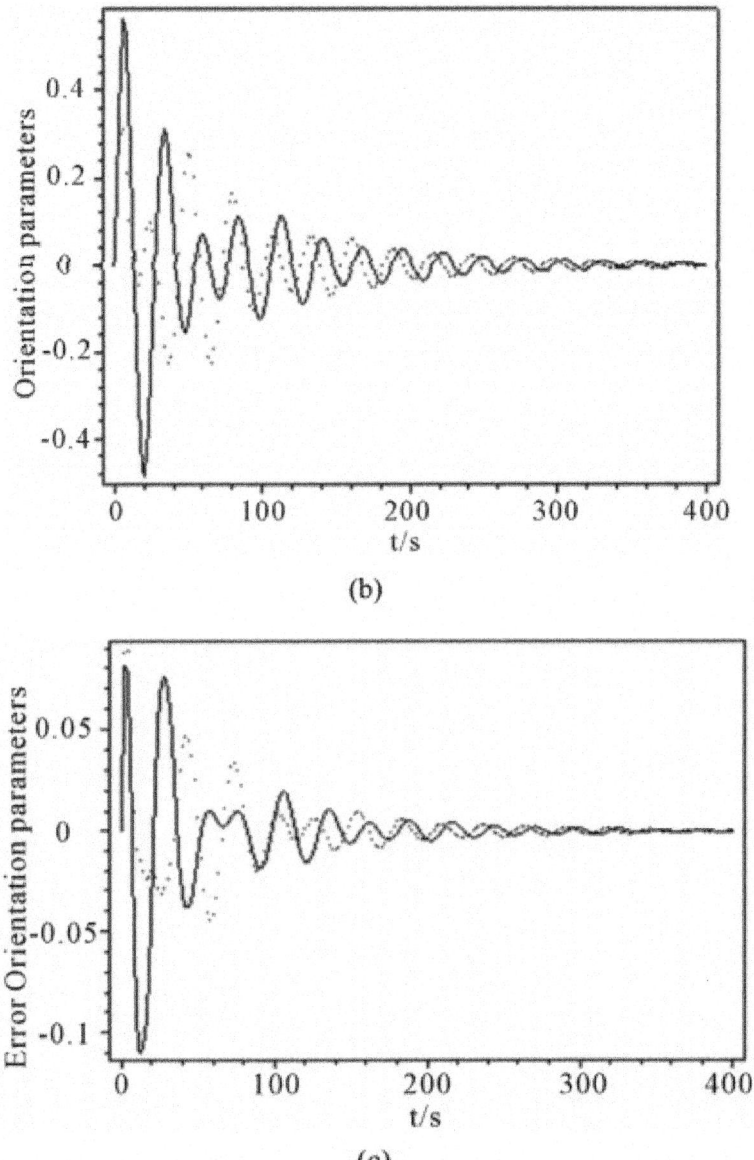

(b)

(c)

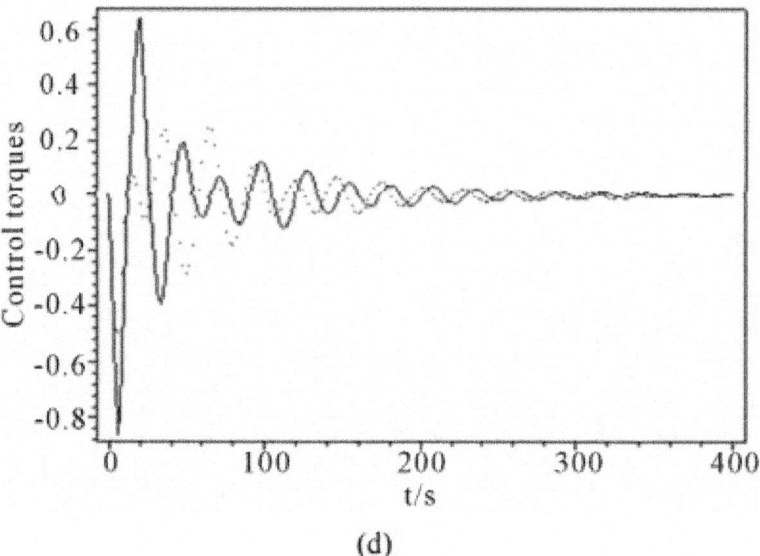

(d)

Figure 2: (a) Body angular velocities; (b) Orientation parameters; (c) Error orientation parameters; (d) Control torques.

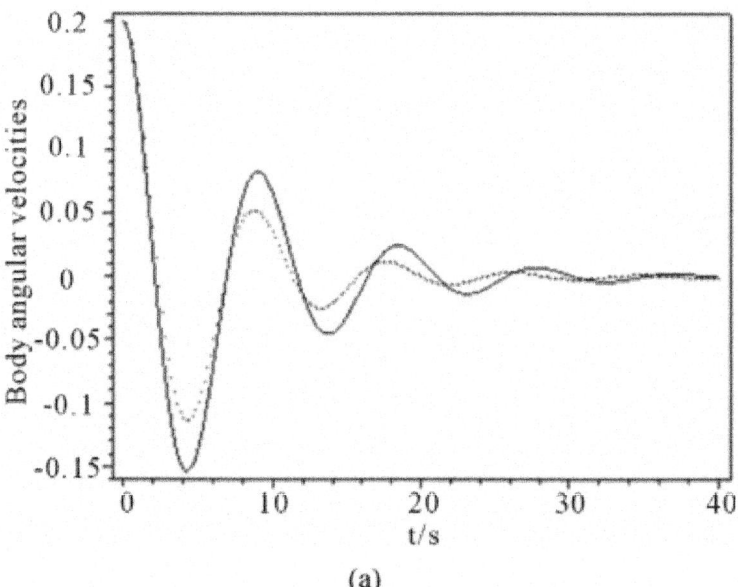

(a)

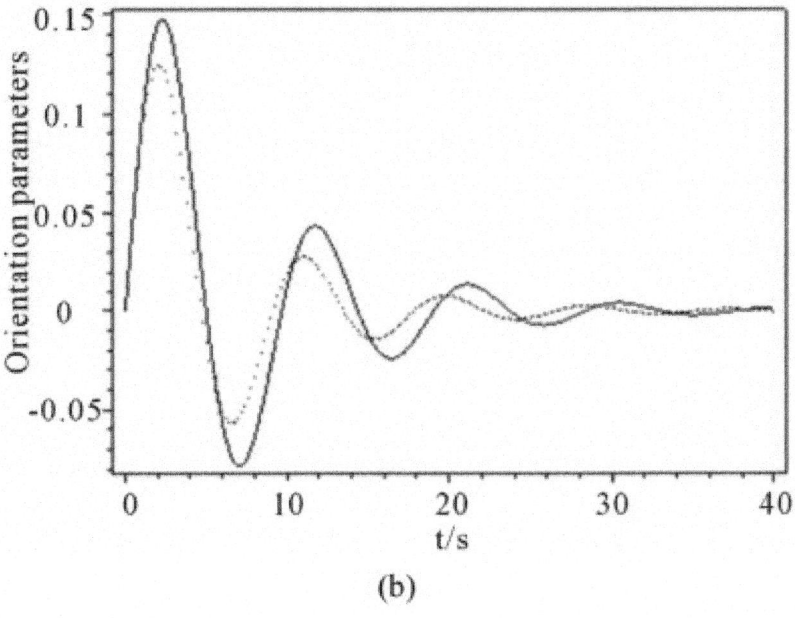

(b)

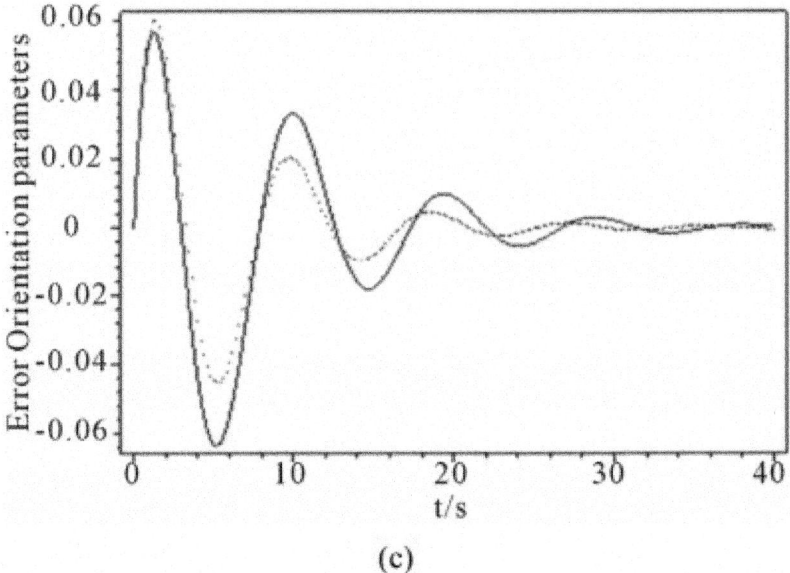

(c)

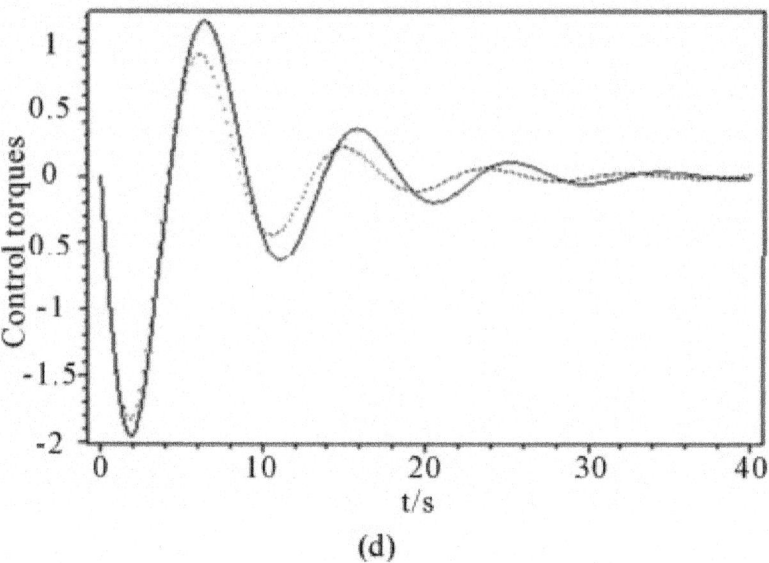

(d)

Figure 3: (a) Body angular velocities; (b) Orientation parameters; (c) Error orientation parameters; (d) Control torques.

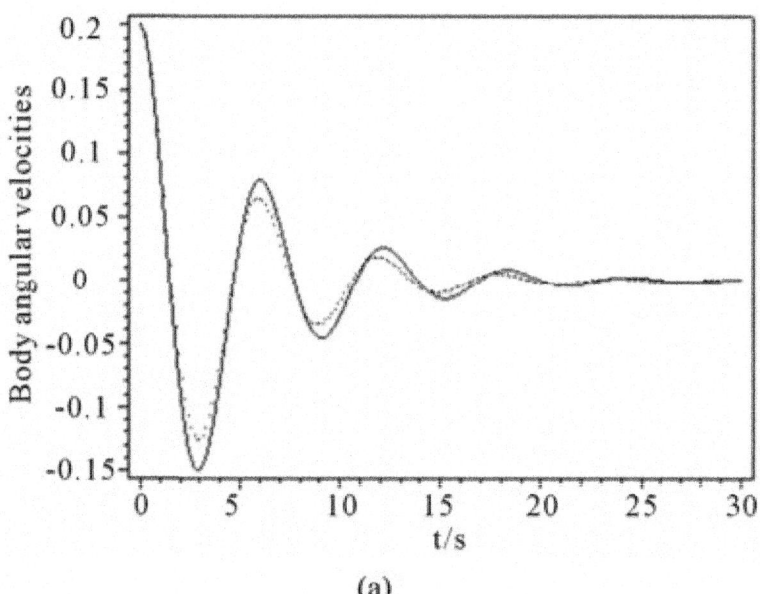

(a)

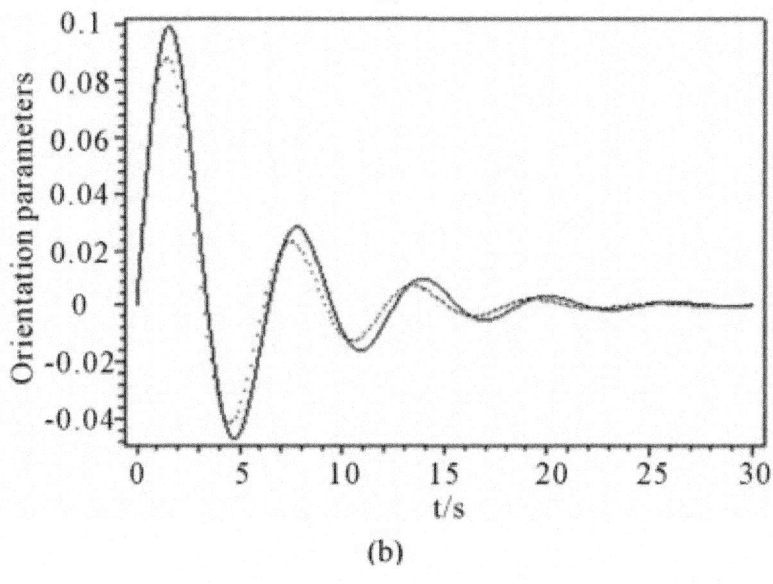

(b)

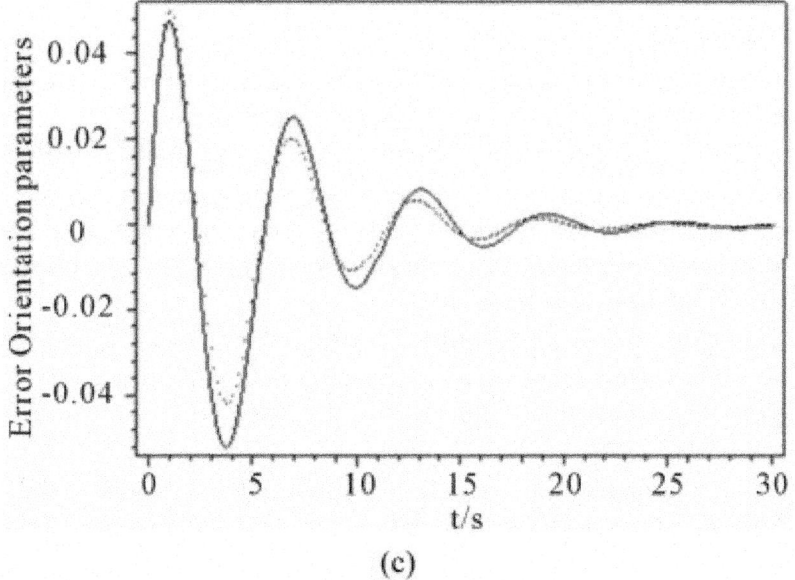

(c)

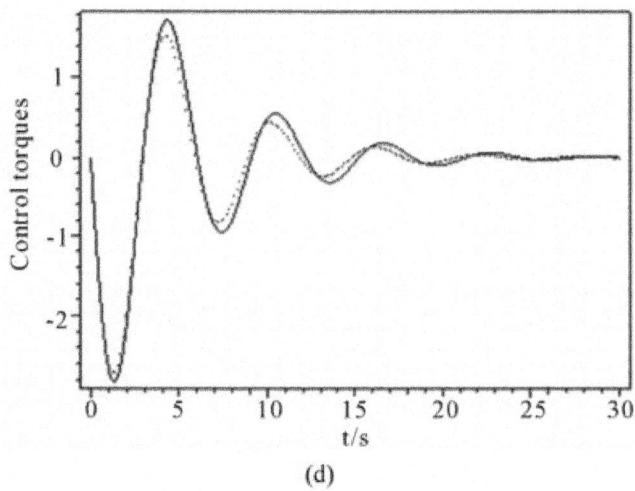

(d)

Figure 4: (a) Body angular velocities; (b) Orientation parameters; (c) Error orientation parameters; (d) Control torques.

REFERENCES

1. A. Abdessam and A. Tayebi, "Global Trajectory Tracking Control of VTOL-UAVs without Linear Velocity Measurements," Automatica, Vol. 46, No. 6, 2010, pp. 1053-1059. doi:10.1016/j.automatica.2010.03.010

2. S. Ding, S. Li and Q. Li, "Adaptive Set Stabilization of the Attitude of a Rigid Spacecraft without Angular Velocity Measurements," Journal of Systems Science and Complexity, Vol. 24, No. 1, 2011, pp. 105-119. doi:10.1007/s11424-011-8214-1

3. A. El-Gohary and T. S. Tawfik, "Attitude Stabilization Using Modified Rodrigues Parameters without Angular Velocity Measurements," World Journal of Mechanics, Vol. 1, No. 2, 2011, pp. 57-63. doi:10.4236/wjm.2011.12008

4. P. Tsiotras and J. M. Longuski, "A New Parameterization the on the Attitude Kinematics," Journal of the Astronautical Sciences, Vol. 43, No. 3, 1995, pp. 243-262.

5. F. Boyer and M. Alamir, "Further Results on Controllability of a Two-Wheeled Satellite," Journal of Guidance, Control, and Dynamics, Vol. 30, No. 2, 2007, pp. 611-619. doi:10.2514/1.21505

6. N. M. Horri and S. Hodgart, "Attitude Stabilization of an under Actuated Satellite Using Two Wheels," IEEE Aerospace Conference, Vol. 6, 2003, pp. 2629-2635.

7.   C. Aguilar-Ibaez, M. S. Suarez-Castanon and F. GuzmanAguilar, "Stabilization of the Angular Velocity of a Rigid Body System Using Two Torques: Energy Matching Condition," American Control Conference, Seattle, 11-13 June 2008, pp. 4845-4849.

8.   H. Shen and P. Tsiotras, "Time-Optimal Control of Axisymmetric Rigid Spacecraft Using Two Controls," Journal of Guidance, Control, and Dynamics, Vol. 22, No. 5, 1999, pp. 682-694. doi:10.2514/2.4436

9.   K. Sungpil and K. Youda, "Spin-Axis Stabilization of a Rigid Spacecraft Using Two Reaction Wheels," Journal of Guidance, Control, and Dynamics, Vol. 24, No. 5, 2001, pp. 1046-1049. doi:10.2514/2.4818

10.   R. Tammepõld, P. Fiorini and M. Kruusmaa, "Attitude Control of Small Hopping Robots for Planetary Exploration: A Feasibility Study," 11th Symposium on Advanced Space Technologies in Robotics and Automation (ASTRA 2011) ESA/ESTEC, Noordwijk, April 2011, pp. 12-14.

11.   P. Tsiotras, "Optimal Regulation and Passivity Results for Axisymmetric Rigid Bodies Using Two Controls," Journal of Guidance, Control, and Dynamics, Vol. 20, No. 3, 1997, pp. 457-463. doi:10.2514/2.4097

12.   H. Khalil, "Nonlinear Systems," MacMillan, Princeton, 1992.

CITATION

CHAPTER 1

Naschie, M. (2014) Cosmic Dark Energy Density from Classical Mechanics and Seemingly Redundant Riemannian Finitely Many Tensor Components of Einstein's General Relativity. *World Journal of Mechanics*, 4, 153-156. doi: 10.4236/wjm.2014.46017.

CHAPTER 2

Silva, A. , Silveira-Neto, A. and Lima, A. (2015) Rotational Oscillation Effect on Flow Characteristics of a Circular Cylinder at Low Reynolds Number. World Journal of Mechanics, 5, 195-209. doi:10.4236/wjm.2015.510019.

CHAPTER 3

RobertoCapata,LeoneMartellucci, (2015) Aerodynamic Brake for Formula Cars. World Journal of Mechanics,05,179-194. doi: 10.4236/wjm.2015.510018

CHAPTER 4

Fung, R. and Lin, C. (2015) Adaptive Real-Coded Genetic Algorithm for Identifying Motor Systems. *Modern Mechanical Engineering*, 5, 69-86. doi: 10.4236/mme.2015.53007

CHAPTER 5

Nabil W.Musa,V. I.Gulyayev,L. V.Shevchuk,HasanAldabas, (2015) Whirl Interaction of a Drill Bit with the Bore-Hole Bottom. *Modern Mechanical Engineering*,05,41-60. doi:10.4236/mme.2015.53005

CHAPTER 6

Kluge, P. , Germaine, D. and Crépin, K. (2015) Dry Friction with Various Frictions Laws: From Wave Modulated Orbit to Stick-Slip Modulated. *Modern Mechanical Engineering*, 5, 28-40. doi: 10.4236/mme.2015.52004.

CHAPTER 7

Denkena, B. , Eckl, M. and Brouwer, D. (2014) Development of a Multiple Degree of Freedom Knee Disarticulation Prosthesis with Active Leg Length Variation. *Modern Mechanical Engineering*, 4, 207-221. doi:10.4236/mme.2014.44020.

CHAPTER 8

Jakkula AnandRao,GandamallaVasumathi,JakkulaMounica, (2015) Joule Heating and Thermal Radiation Effects on MHD Boundary Layer Flow of a Nanofluid over an Exponentially Stretching Sheet in a Porous Medium. *World Journal of Mechanics*,05,151-164. doi: 10.4236/wjm.2015.59016

CHAPTER 9

Tom Joyce, Iain Evans, William Pallan & Clare Hopkins (2013) A Hands-on Project-based Mechanical Engineering Design Module Focusing on Sustainability, Engineering Education, 8:1, 65-80

CHAPTER 10

Liu, Z. and Chen, H. (2016) Mechanical Design and Kinematic Simulation of Automated Assembly System for Relay. *World Journal of Mechanics*, 6, 1-7. doi: 10.4236/wjm.2016.61001.

CHAPTER 11

W. Elsamee, "Evaluation of the Ultimate Capacity of Friction Piles," *Engineering*, Vol. 4 No. 11, 2012, pp. 778-789. doi: 10.4236/eng.2012.411100.

CHAPTER 12

T. Tawfik, "Attitude Control of an Axi-Symmetric Rigid Body Using Two Controls without Angular Velocity Measurements Paper," *World Journal of Mechanics*, Vol. 3 No. 5A, 2013, pp. 1-5. doi:10.4236/wjm.2013.35A001.

INDEX